Numerical Methods in Environmental Data Analysis

Numerical Methods in Environmental Data Analysis

Moses Eterigho Emetere

*Department of Mechanical Engineering Science,
University of Johannesburg, South Africa
Department of Physics, Covenant University, Ota, Ogun, Nigeria*

ELSEVIER

Elsevier
Radarweg 29, PO Box 211, 1000 AE Amsterdam, Netherlands
The Boulevard, Langford Lane, Kidlington, Oxford OX5 1GB, United Kingdom
50 Hampshire Street, 5th Floor, Cambridge, MA 02139, United States

Notices

Knowledge and best practice in this field are constantly changing. As new research and
experience broaden our understanding, changes in research methods, professional
practices, or medical treatment may become necessary.

Practitioners and researchers must always rely on their own experience and knowledge in
evaluating and using any information, methods, compounds, or experiments described
herein. In using such information or methods they should be mindful of their own safety
and the safety of others, including parties for whom they have a professional responsibility.

To the fullest extent of the law, neither the Publisher nor the authors, contributors, or
editors, assume any liability for any injury and/or damage to persons or property as a
matter of products liability, negligence or otherwise, or from any use or operation of any
methods, products, instructions, or ideas contained in the material herein.

ISBN: 978-0-12-818971-9

For information on all Elsevier publications visit our website at
https://www.elsevier.com/books-and-journals

Publisher: Candice G. Janco
Acquisitions Editor: Peter Llewellyn
Editorial Project Manager: Aleksandra Packowska
Production Project Manager: Sreejith Viswanathan
Cover Designer: Mark Rogers

Typeset by TNQ Technologies

Contents

Preface ix

CHAPTER 1   Overview on data treatment....................................... 1
   1 Introduction 1
      1.1 Mathematical technique...6
      1.2 Computational technique...7
      1.3 Statistical data treatment 7
   References...11

CHAPTER 2   Case study in environmental pollution research 13
   1 Introduction 13
      1.1 Air pollution... 14
      1.2 Land pollution... 21
      1.3 Water pollution.. 24
      1.4 Noise pollution.. 32
      1.5 Radioactive pollution.. 33
      1.6 Electronic waste pollution 35
   References...38
   Further reading...39

CHAPTER 3   Typical environmental challenges...........................41
   1 Introduction 41
      1.1 Thermal comfort as a source of environmental concern.......41
      1.2 Rainfall as a source of environmental concern 44
      1.3 Recent environmental crisis and the problem of climate
          change 47
   References...51

**CHAPTER 4   Generating environmental data: Progress and
                   shortcoming**...53
   1 Method of generating environmental data: common
      challenges, safety, and errors...53
      1.1 Data quality and errors 55
      1.2 Satellite measurement.. 60
      1.3 Modeling procedure 63
      1.4 Experimental procedure 69
   2 Common errors in laboratory practice....................................74
   3 Maintaining laboratory apparatus..75
   References...76

CHAPTER 5   Root finding technique in environmental research...79
   1 Application of root finding technique to environmental
     data 79
     1.1 The root finding method..79
     1.2 Modification of the root finding method to data
        application..82
     1.3 Computational application of root finding method to
        data application 103
     Reference...117

CHAPTER 6   Numerical differential analysis in environmental
               research...119
   1 Introduction 119
     1.1 Euler method 121
     1.2 Improved Euler method 122
     1.3 Runge−Kutta method..123
     1.4 Predictor Corrector method.......................................126
     1.5 Midpoint method 128
     1.6 Application of numerical methods of solving
        differentiation in environmental research.......................128
     1.7 Computational processing of numerical methods
        for solving differential equation.................................136
     1.8 Computational application of derivatives to
        environmental data...142
     1.9 Case 1: derivative of experimental data........................142
     References..147
     Further reading..148

CHAPTER 7   Numerical integration application to
               environmental data 149
   1 Introduction 149
     1.1 Midpoint..149
     1.2 Trapezoidal rule...151
     1.3 Simpson's rule...154
     1.4 Computational application of numerical integration 158
     References..168

CHAPTER 8   Numerical interpolation in environmental
               research..169
   1 Introduction 169
   2 Application of interpolation to environmental data.................170
   3 Lagrange interpolation 172

4 Newton interpolation 176

5 Spline interpolation 179

6 Computational application of interpolation 181

References.. 189

**CHAPTER 9 Environmental/atmospheric numerical models
formulations: model review 191**

1 Introduction 191

1.1 Global forecast system... 191

1.2 NOGAPS-ALPHA model 192

1.3 Global Environmental Multiscale Model (GEM)............. 195

1.4 European Center for Medium Range Weather Forecasts... 196

1.5 Unified Model (UKMO)... 197

1.6 French global atmospheric forecast model (ARPEGE)..... 199

1.7 Weather Research and Forecasting (WRF)..................... 200

1.8 Japan Meteorological Agency Nonhydrostatic Model
(JMA-NHM) 203

1.9 The fifth generation mesoscale model........................... 205

1.10 Advanced Region Prediction System (ARPS)................. 206

1.11 High Resolution Limited Area Model (HIRLAM)........... 207

1.12 Global Environmental Multiscale limited area model 208

1.13 ALADIN model.. 210

1.14 Eta model 213

1.15 Microscale model (MIMO) 215

1.16 Regional atmospheric modeling system (RAMS)............ 216

References.. 217

Further reading... 221

Index.. 223

Preface

Environmental data may be described in terms of quantitative, qualitative, or geographically referenced facts that represent the state of the environment and its changes. Quantitative environmental data consist of data, statistics and indicators of databases, spreadsheets, compendia, and yearbook type products. Qualitative environment data are descriptions (e.g., textual, pictorial) of the environment or its constituent parts that cannot be adequately represented by accurate quantitative or geographically referenced descriptors. Geographically referenced environmental data are described in digital maps, satellite imagery, and other sources linked to a location or map feature. Summarily, it can be postulated that dataset in environmental studies is like blood to the human body. All decisions in environmental studies are based on observables that are measurable, reliable, realistic, and consistent with theories. Environmental theories are formulated from observables. Hence, a faulty observable can lead to a colossal failure in processes, prediction, model formulation, and decision.

The inevitable outcomes of climate change have redefined observables such that new theories and models are necessary due to data inconsistency, noise, and spikes. Aside from just getting dataset and simulating, it is now expedient that the integrity of a dataset be the first line of operation in data analytics. This feat can be achieved through the guidance of proven theories. The knowledge of this theory, when to apply it on a dataset, how to apply it, and ways to validate emerging results are salient in any field of environmental sciences. Hence, the focus of this book is to educate beginners and professionals on the above.

Environmental indicators are usually the environment statistics that are in need of further processing and interpretation. Based on this, there is the need of the application of numerical methods to validate, expatiate, predict, back-trace, and create new possibilities. Validation technique through numerical methods enables the researcher to ascertain the pattern trend of series of observables and tie them to certain established theories. Expatiation technique through numerical methods enables the researcher to take an informed numerical guess to replace missing data, noise, and data anomalies. Missing data is common in atmospheric research. Missing data makes the genuity of the data to be questionable especially when the user is a beginner or novice. Assume if the satellite measurement of a parameter shows missing values for 7 months in a yearly dataset. Ignoring the missing data for the remaining 5 months would certainly be erroneous to analyze monthly or seasonally. The same scenario applies to noise in data and data anomalies. This book seeks to train beginners and professionals on the aforementioned expertise.

Overview on data treatment

1. Introduction

Data is usually defined as raw or unprocessed facts or statistics that will need to be processed or interpreted in order to get information. Technically, there are three types of data based on their source and availability: primary, secondary, and mosaic. Primary data is data that is collected through firsthand experiences, studies, or research. Secondary data is data or information that has been collected from other sources. Mosaic data refers to data and information that is collected by putting together bits and pieces of information that are already publicly available. Environmental data are large amounts of unprocessed observations and measurements about the environment (or its components) and related processes. Data used for the production of environment output, report, or statistics are compiled by many different collection techniques and institutions whose data sources are hosted privately or publicly at known sites. Understanding and knowing the pros and cons of each source is key in environment reportage. Data sources are the initial locations where the collected data originates from and runs public object for the establishment and can be a flat file, database, scraped web data, social media, and database access which profuse across the internet. Data source is considered to help users and applications to secure and move data to where it needs to be. The purpose of the data source is to bundle connection information that is easier to comprehend. In environmental science, data source can be classified into two: the primary and secondary data. The primary data is original and accurate and is collected with the aim of getting the solution to a problem at hand, and it includes surveys, observations, websites, questionnaires, etc. It is reliable, objective, and authentic. The secondary data are data that are readily available and are more accesible to the public than the primary data (e.g., industry surveys, compilation).

The type of data that could be obtained from research could either be qualitative or quantitative. Qualitative data research centers around getting information concerning the attribute, characteristics, or qualities of sample. It does not involve numbers. While quantitative data research are research studies whose data are quantifiable with the use of numbers, where data are computed through discrete whole number integers or continuous floating point values. There are a lot of examples of numerical data; however, they are all categorized into two types: discrete and

Numerical Methods in Environmental Data Analysis. https://doi.org/10.1016/B978-0-12-818971-9.00001-6

continuous data. Discrete data are data that take numerical symbols as they are countable list of items. They take values that can be grouped into categories or list, where the list may either be finite or infinite. Discrete data takes number counting from 1 to 10, or 1 to infinity, but it always occurs in a range. Continuous data is a type of numerical data which represents measurements. These data are described as values that take interval such as averages, largest or smallest number (among ranges), and cumulative grade point.

There are different types of data source. Flat file is a database that stores data in a plain text format and teaches how to upload, prepare, and update your csv files to data-pines. This consists of a single table of data types table and cannot contain multiple tables of data types, and it has no folders or paths related to them and is used to import data and store table information. Examples of flat file include plain text, binary file, delimited file, and flat file database. Another type of data source is database. Database is one of the oldest data sources and the relational database is one of the common databases that can easily be connected to the data-pines. Then each database will then be represented as an individual data connection. They support the manipulation of data and electronic storage. The types of database are network database, hierarchical database, and object-oriented database. A typical example of environmental organizations that make use of the flat files is the NASA-associated satellites extension such as MERRA and GIOVANNI. Fig. 1.1 shows the Global Precipitation Measurement (GPM) constellations that have some of their dataset as flat file.

Web Services is a type of data source. It is a system of communication between two electronic devices over a network and is also an assembly of the segment that the software makes available over the internet. And it is formulated to communicate with different programs rather than the users. In a web service the web technology known as the "Http" this data source is used for transmitting machine-readable file format (e.g., the XML). The types of web services include web template, web service flow language, web service conversation language, web service metadata language, and web service description language. Australian department of agriculture, water, and the environment have several web services where a list of environmental data can be downloaded.

The most popular form of data source is databases. Popular environment databases include Proquest Natural Sciences Database, Engineering Village, Green-FILE, Environmental Impact Statement (EIS) Database (EPA), Health & Environmental Research Online, etc. There are several different types of databases, and various companies sell databases with various plans and features. MS Access, Oracle, DB2, Informix, SQL, MySQL, Amazon Simple DB, and a variety of other databases are widely used today. In general, contingent databases—that is, databases that document a company's consistent transactions, such as CRM, HRM, and ERP— are not considered to be suitable for business records. This is attributable to a number of reasons, including the fact that data is not enhanced for itemizing and inspecting, and specifically querying these databases may block the layout and prevent the databases from correctly tracking trades. Organizations can use an ETL tool to

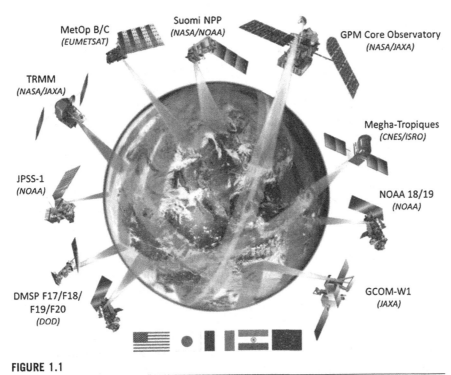

FIGURE 1.1

Flat file user: Global Precipitation Measurement (GPM) constellations (Laviola et al., 2020).

obtain information from their constrained servers, transform it into BI-ready format, and weigh it into a data storage room and perhaps another data store. The one flaw in this theory is that a data circulation focus is a perplexing and expensive plan, which is why many organizations want to report explicitly against their stringent databases.

Online media information is a source of data. It is gathered from long range interpersonal communication administrations like Facebook, microblogging stages like twitter, media sharing destinations like YouTube and Instagram, sites, conversation discussions, client audit locales, and new locales. This information can be gathered from things had been posted, as, acknowledge or search about through your gadgets.

The method of generating primary data in disciplines related to environmental science may be through survey, experiment, and observation. Survey is carried out by questioning individuals based on different topics and reporting their responses, and are used to test the different concepts, reflect the attitude of different people, reporting certain personalities of people, testing hypotheses of people's nature of relationships and personalities. Experiment is an organized study where the analyzer gets to understand the effects, causes, and processes involved in a particular process and involves manipulating one variable to determine if there are changes in the other. The types of experimental design include completely random design, randomized block design, Latin square design, and factorial design etc. Observation is a method that engages vision as it main means of data collection, and is also studying others' behaviors without taking control of it. There are a few things to keep in mind when carrying-out experiment in environmental science:

a. Measurement technique: This technique is relevant because it has an impact on the success of your data. The configuration of the equipment as well as the use of updated standards are essential parameters before taking measurement. Also, the procedures for obtaining live samples are salient in experimental technique.
b. Multiple trials: This includes going through the investigation again and again. The more preliminary work you do, the higher your average value would be and the more accurate and reliable the results would look like.

The method of generating secondary dataset includes internet sources, external sources, satellite measurement etc. Internal sources are dataset that are within the organization and can be obtained within a short effort, a period of time than the external sources and they include internal experts, data mining, sales-force report, miscellaneous report, accounting sources etc. External sources are dataset that are outside the organization and are quite difficult because they have many collections and the sources are much more frequent, and they include syndicate service, governmental publications, nongovernmental publications, etc.

Data treatment is a very essential part of any experimental work or analysis of a secondary dataset. It is essential in all experiments, spanning from scientific to social to business to medicine etc. Data treatment helps researchers identify errors, spot trends, observe correlation and relationships, make inferences, and draw meaning and conclusions from collected data. It involves all the actions and processes in the investigation and collection of data and the additional processes performed on data in order to arrive at useful information, so as to make deductions and inferences. Every environmental researcher, regardless of their field, must have the basic concept of data treatment for their research or their study to be reliable. Data treatment is essential and equally important, as well as data organization, to draw appropriate conclusions in a given data set. Data treatment is a process to ensure its reliability and uniqueness in experiments and data collection designs. This process is vital to efficiently make use of a given data in the right way. It is essential to correctly treat data to maintain the research's authenticity, accuracy, and reliability. A well-defined understanding is needed to perform suitable experiments with the correct information obtained from any given data set. Data treatment can be descriptive, that is, describing the relationship between variables in a population set so as to distinguish between a noise, spike, and trend. It can also be inferential, that is, testing a given hypothesis by making inferences from a collected data set or an establish law or theory. To obtain the desired result, data must be processed using a variety of methods. All experiments randomly produce errors or noise. Data noise can either be systematic or random errors. It is advisable that errors and noise be taken into consideration in the course of the experiment for the result of the experiment to make sense.

Regardless of how cautious a researcher can be while measuring or extracting samples in an environment, all experiments are vulnerable to inaccuracies caused by three forms of errors: systematic, random, and spontaneous errors. These errors are most times spotted during the treatment of data, and the correction can then be reintegrated in the process. Spontaneous errors are widely reported in genetic code (Griffiths et al., 2000). Systematic errors are errors that are caused by either the data collection equipment or the method used to collect the data. Internal error can emerge from measuring or characterizing instruments which most of the time possess random errors that occur accidentally or unpredictably in the experimental configuration. This type of error will continue to occur in all instances of the experiment until the source of the error is addressed. Some examples of this kind of error are an incorrectly calibrated measuring device, a worn out instrument, and a misconception on the observer's end. Systematic errors are usually consistent in the amount of error in the measured value. These experimental errors can lead to two different kinds of conclusion errors: type 1 and type 2 errors. A type 1 error occurs when a researcher rejects a true null hypothesis, resulting in a false positive. A type 2 error, on the other hand, is a false negative caused by a researcher's inability to reject a false null hypothesis. In other words, the method of data treatment employed in research depends on the field of research or kind of experiment being conducted, as this would affect the kind of data being collected, and the desired form of the data required to arrive at a conclusion. Random errors are errors that are caused by irregular and unpredictable variations in the experiments. This variation could be as a result of external environmental conditions surrounding the experiment; it could also be caused by a fault in the measuring instrument. These types of errors do not usually have the same errors in the same direction for all instances of the experiment. Random errors occur unknowingly or unpredictably in the experimental configuration. They arise unknowingly or unpredictably in the experimental setup.

Data treatment is one of the last operations in data analytics. There is preceding operation i.e., data collection, data preparation, data processing, data cleaning, etc., that must be done before data treatment. Data collection is one of the initial stages of every research endeavor that involves the collecting of data from all available platforms. This could be through surveys and experiments in the laboratory or sites. It is required that the data source be relevant, reliable, and authentic. Data preparation is the process that often follows after the data collection stage. The data preparation stage is often referred to as pre-processing stage. This is the stage at which data is organized before it is processed into the required form. Data processing is the stage at which data is translated into the readable, relatable, and required format. It might involve placing data into rows and columns, and it might require the use of a computer to process the input data. It may require complex programming, algorithms, etc. The method of processing of data depends on the type of data to be processed, processing tool/software, and size of dataset. For example, for a small ASCII dataset and Microsoft excel are commonly used. For big ASCII data, structured programming language is used to save time and reduce errors. Data cleaning is the process where noise in data are removed. It is synonymous to data treatment but it is the preliminary stage before data treatment. For example, when datasets are downloaded from a satellite station in ASCII format, there could be missing data, which most of the time appear as "9.9999," "***," and "9999" or blank. The removal of this anomalies is data cleaning not data treatment. Also, in the data cleaning stage, unnecessary data can be removed such as duplicates and errors. The data cleaning process involves deduplication, matching records, identifying data inconsistencies, checking the overall data quality, etc. The emerging dataset after data cleaning is expected to be in the required, readable format. This readable format could be in the form of an equation, image, video, graph, theory etc. The information obtained from this stage is what will be used for data treatment.

There are three ways of data treatment in literature. They are:

(a) Mathematical technique (statistical data treatment)
(b) Computational technique (algorithm data analysis)
(c) Statistical technique

1.1 Mathematical technique

This is a technique that involves the use of mathematical theories, formulae, and mathematical manipulation. Some of these mathematical processes include:

I. Regression analysis: This is an analysis used to evaluate the relationship between two or more set of numerical data. When using this technique, we look for a correlation between the dependent numerical data and any number of independent variables that might have an effect on these numerical data. The aim of regression analysis is to estimate how one or more variables might impact the dependent numerical data, in order to identify trends and patterns. This was used specifically for prediction and forecasting future trends. It is also important to note that regression analysis only helps to determine whether or not there is a relationship between a set of numerical set of data, and it does not say anything about the cause or effect.

II. Factor analysis: This is a technique used to reduce a large set of variables to a smaller number of variables. It works on the idea of multiple separate, observable variables correlate with each other because they are all associated with an underlying set. This is useful not only because it reduces variable in a particular set of numerical data into smaller understandable variables, but it also helps to uncover hidden patterns.

III. Time series analysis: This is a statistical technique used to identify numerical data using time interval. It records and separate data into groups based on the data that have similar time interval or the time created.

Numerical analysis is mostly needed to solve engineering problems that result into equations that cannot be solved analytically with simple formulas. Some applications are listed here:

a. Modern applications and computer software: Most sophisticated numerical analysis software is embedded in popular software packages, e.g., spreadsheet programs.

b. Business applications: Modern businesses these days make much use of optimization methods in deciding what or how to allocate a resource most efficiently, such as inventory control, scheduling, budgeting, and investment strategies.

1.2 Computational technique

This is a technique that involves the use of AI systems such as the computer system. This involves using programmed codes, encoded scripts formulas to arrange and present numerical data in an organized manner meaningful to interpret and use. There are a lot of programming software created to solve this problem. Some of the best ones include these:

I. Analytica: This is a software created and developed by Lumina Decision Systems for receiving/retrieving, analyzing, and communicating numerical data. It uses hierarchical influence diagrams for visual creation and view of models, intelligent arrays for working on multidimensional data.

II. MATLAB: Matrix Laboratory is a proprietary multi-paradigm programming language and numeric computing working environment developed by MathWorks. MATLAB makes it possible for matrix manipulations, plotting of functions and data, implementation of algorithms, creation of user interfaces, and interfacing with programs written in other languages. MATLAB is made for the source purpose of numerical data treatment.

III. FlexPro: This is a software designed for the analysis and presentation of scientific and technical data. This software was created by the Weisang GmbH team. It was designed to run Microsoft windows. FlexPro can analyze large amount of data with high sampling rates. All data to be analyzed are stored in an object database. FlexPro has a built-in programming language, FPScript, which is optimized to carry out data analysis and support direct operations on non-scalar objects such as vectors and matrices as well as composed data structures like signal series.

IV. FreeMat: A free open-source numerical data treatment environment and programming language, similar to MATLAB.

V. jLab: This is a numerical computational environment created with a Java software and interface.

1.3 Statistical data treatment

There are various methods involved in the treatment of data, and one of the most common methods is the statistical method of treatment of data. When you apply a statistical approach to a data set in order to turn it from a list of meaningless numbers into useful output, this is known as the statistical treatment of data. Statistical method includes but not limited to; mean, median mode, range, standard deviation, conditional probability, range, distribution range, sampling, correlation, regression, and probability. There are some notable errors in data treatment, and using statistical techniques to classify potential outliers and errors is an important aspect of data processing. Statistical data treatment is one of the essential aspects of any experiment conducted today. It can be seen using any known statistical method to draw meaning from a set of given meaningless data sets. Statistical distribution can be classified into two groups. To begin with, one of them is considered to have discrete random variables, which means that each word includes a single numerical value. The second form of statistical distribution, which includes continuous random variables, is called a continuous random variable distribution (the data is known to take infinitely many values). Statistical data treatment often entails defining the data collection, and one of the most effective ways to do so is to use the measure of core tendencies such as the mean, mode, and median.

The core tendencies described above make it simple for any researcher to perform any research experiment and understand how the data set is concentrated. Central tendencies such as the standard deviation, range, and uncertainty help the researcher understand the data set's distribution. Nevertheless, care should consistently be taken to assume that all data sets are the same and evenly distributed. Any of the above-mentioned central tendencies can be used to ensure that.

This method involves using some statistical methods to transform a given meaningless data into meaningful data sets. It involves the use of some statistical methods:

➤ MEAN: In statistics, this is a key idea. It describes the characteristics of a statistical distribution. In a set of numbers, it is the most common value.

To measure it, take into account the figures of the relative multitude of terms and then divide by the number of terms. The mean of a collection of data can be determined in several ways. It can be determined using the arithmetic mean process, which involves dividing the total number of data sets by the sum of the total number of data sets. To find the mean, add all of the numbers in a set together, then divide the total by the total number of numbers. A dataset's mean can also be calculated by a method known as the geometric mean, which is the nth root of the product of all numbers in the data set. It includes the volatility and compounding effects of returns. The arithmetic mean, also known as the mean or standard, is the sum of a set of values divided by the number of values in the group.

$$\text{Mean} = \frac{\text{Sum of all data points}}{\text{Number of data points}}$$

➤ MODE: The estimate of the word that occurs often in the form of dissemination with a discrete arbitrary variable. The mode is the number that happens frequently inside a bunch of numbers. It is feasible to have two modes (bimodal), three modes (trimodal), or more modes inside bigger arrangements of numbers. Bimodal appropriation refers to the appropriation that has two modes. Trimodal appropriation is a three-mode appropriation. The most severe estimate of capability is the form of dispersion with a constant irregular variable. Similarly, discrete appropriations can have more than one mode. In this case, it takes special expertise to identify errors or noise in a given dataset. The advantages of the mode is its simplicity to identify and determine a value. Its disadvantage of mode is the possibility that a set of values might have only one mode, or no mode at all. Also, mode is not stable when the statistics has small numbers.

➤ RANGE: The range of your data in statistics is the range from the lowest to the highest value of the distribution. By subtracting the lowest from the highest value, the spectrum is determined. A wide variance in a distribution implies high variability, while a small range indicates low variability.

➤ MEDIAN: The middle value in distribution is referred to as the arithmetic median, which is a positional average. It divides the sequence into two halves by grouping the elements in ascending or descending order of magnitude before finding the middle value and is denoted by the symbol X or M. It can also be referred to as the middle position or the middle class, or median class. For example, in a set of numbers 1, 2, 3, 4, 5, the median of the set of numbers would be 3. In the case where two number are in the middle class (e.g., 1, 2, 3, 4, 5, 6) the median of the set of numbers is the average of 3 and 4 which is 3.5.

➤ STANDARD DEVIATION: The standard deviation is a calculation of a group of values' variance or dispersion. A low standard deviation means that the values are similar to the set's mean (also known as the predicted value), while a high standard deviation indicates that the values are distributed out across a greater spectrum. This can be calculated with the formula

$$\sigma = \sqrt{\frac{\sum (x_i - \mu)^2}{N}}$$

where

σ = standard deviation
N = size of the population
x_i = each value from the population
μ = population mean

➤ SAMPLING: Data sampling is a predictive research methodology that involves selecting, manipulating, and analyzing a representative subset of data points in order to uncover correlations and trends in a broader data collection. There are various methods used to sample data:
- Simple random sampling
- Systematic sampling
- Stratified sampling
- Cluster sampling

Sampling is a method of statistical surveying in which a predetermined number of observations are taken from a larger group of individuals. The method used to collect data from a larger group of individuals varies depending on the type of study being conducted, but it can include basic discretionary sampling or precise sampling.

Sampling is the selection of a sample of patients from within a measurable population to determine the population's attributes. Sampling is a realistic approach that is concerned with the individual's interpretation preference.

➢ CONDITIONAL PROBABILITY: The probability of one occurrence happening in the context of one or more other events is known as conditional probability. Conditional probability denotes the likelihood of a certain outcome occurring if another event has already happened. It is always expressed as the probability of B given an A, and it is written as $P(B|A)$, where the probability of B is an infinite supply of events.

➢ DISTRIBUTION RANGE: The range of a species is the geographical area within which that species can be found. Within that range, distribution is the general structure of the species population, while dispersion is the variation in its population density. The range is the smallest stretch that includes all the details and has a touch of measurable displacement. It is rated in the same units as the data. It is generally helpful in contributing to the dispersion of small instructional assortments, and it has an infinite supply of discernments.

➢ REGRESSION: Regression is a mathematical technique used in economics, investing, and other fields to evaluate the intensity and nature of a relationship between one dependent variable (usually denoted by Y) and a set of other variables (known as independent variables):

$$Y_i = f(X_i, \beta) + e_i$$

where

Y_i = dependent variable
f = function
X_i = independent variable
β = unknown parameters
e_i = error terms

Three major uses of regression analysis are:
• Determining the strength of predictors
• Predicting an effect
• Trend forecasting

Types of regression include:
- Linear regression
- Polynomial regression
- Ridge regression
- Lasso regression
- Elastic net regression

This method includes several variations such as linear and multiple linear. Regression analysis offers numerous applications in various disciplines, including finance. Linear regression is based on six fundamental assumptions: the dependent and independent variables show a linear relationship between the slope and the intercept; the independent variable is not random; the value of residual error is zero; the value of the residual error is constant across all observations; the value of the residual error is not correlated across all observations; the residual error values follow the normal distribution. Linear regression is a module that assesses relationship between a dependent variable and an independent variable.

Multiple linear regression is similar to the simple linear regression in a way, with the exceptions of the multiple independent variables are use in the model. Noncollinearity-multiple variables should show a minimum of correlation with each other. If the independent variables are highly correlated with each other, it will be difficult to access the true relationship between the dependent and independent variables.

➤ VARIANCE: A statistical calculation of the spread between numbers in a data set is known as a variance. Variance quantifies how far each number in the set deviates from the mean, and hence from any other number in the set.

References

Griffiths, A.J.F., Miller, J.H., Suzuki, D.T., et al., 2000. An Introduction to Genetic Analysis, seventh ed. W. H. Freeman, New York. Spontaneous mutations. Available from: https://www.ncbi.nlm.nih.gov/books/NBK21897/.

Laviola, S., Monte, G., Levizzani, V., Ferraro, R.R., Beauchamp, J., 2020. A new method for hail detection from the GPM constellation: a prospect for a global hailstorm climatology. Rem. Sens. 12 (21), 3553. https://doi.org/10.3390/rs12213553.

Case study in environmental pollution research

1. Introduction

Pollution can be defined as the addition of hazardous and toxic materials to the environment, thereby causing adverse effects. Pollution can also be defined as the introduction of pollutants which could be in the form of harmful substances or energy into the atmosphere which eventually becomes detrimental to it. There are three main types of pollution:

- Air pollution
- Water pollution
- Land pollution

However, there are other equally important types of pollution such as noise pollution, thermal (heat) pollution, plastic pollution, radioactive pollution, and light pollution. A pollutant is any material or substance that contains properties that are harmful to the biotic and abiotic system. Pollutants can simply be defined as constituents that make up or are involved in pollution. They are the main composition of pollution. Pollutants can be divided into two categories:

1. Primary pollutants
2. Secondary pollutants

Primary pollutants are the pollutants at the first point of introduction into the environment, while the secondary pollutants are the pollutants that are formed from these primary pollutants and the adverse effects of other external factors on them. Pollutants can be of any form whether solid, gaseous or liquid, or radioactive, sound and heat energy. Pollutants are mostly anthropogenic (man-made pollutants), but in some cases, pollution can be caused by natural events such as wildfires, where the air is contaminated.

Pollution is as old as mankind since the ancient times before civilization—from the fires they created to the waste they left behind. Although it was not a matter of great concern at the time, with the increase in population, the quick spread of industrialization and civilization and establishments of towns and cities, pollution is now an issue that proposes danger in the years ahead. With the increase of environmental

Numerical Methods in Environmental Data Analysis. https://doi.org/10.1016/B978-0-12-818971-9.00003-X

pollution and pollutants, efforts have been made to provide awareness to countries, states, and towns, and laws have been passed to reduce pollution and control the damage that has already been done to the environment. Some of these laws are

- The Air Pollution Control Act of 1955, United States
- Biological Diversity Act of 2002, India
- Oil Pollution of The Sea (Civil Liability and Compensation) (Amendment) Act of 2003, Ireland
- Environmental (Prevention of Pollution in Coastal Zone and Other Segments of The Environment) Regulation of 2003, Kenya
- Clean Water Act of 1972, United States
- Clean Air Act of 1970, United States
- Pollution Prevention Act of 1990, United States
- Environmental Management Act of 1997, Netherlands
- Pollution Control Act of 1981, Norway

1.1 Air pollution

According to World Health Organization (WHO), 9 out of 10 people breathe highly contaminated air. Air pollution can be defined as the presence or addition of harmful particulates (such as aerosols) or gases (such as greenhouse gases) to the atmosphere that are detrimental to the well-being of human beings and other living organisms and cause damage to the ozone and climate. Some examples of these harmful substances include chlorofluorocarbon (CFC), ammonia, nitrogen oxide (NO_x), carbon monoxide (CO), exhaust fumes (soot) etc. Air pollution can be classified under indoor and outdoor air pollution.

Air pollution is one of the biggest risk factors in the world as it causes up to 5 million deaths each year and is the cause of 9% of deaths around the world. In some developed countries, death rates have been on a decline due to the control and reduction measures of indoor air pollution such as improving proper ventilation, reducing the use of a fireplace. Also, the reduction of outdoor pollution through the enactment of laws and decrees that has strict implications on industrial emissions, anthropogenic emissions, and emissions from unconventional sources such as sewage. The unconventional sources are the new area of research as it is found to emit dangerous bioaerosols into the environment. Most of the bioaerosols are pathogenic. The anthropogenic emissions is the most common, and it can appear as one of the following.

Burning of fossil fuels: Most of the air pollution takes place due to the burning of fossil fuels. Over the years, the burning of fossil fuels has been almost inevitable because fossil fuels have been one of the major sources of energy, electricity, and power generation. In the United States, fossil fuel consumption has nearly tripled within the last 50 years. When these fuels are burnt, they release harmful gases such as carbon monoxide, i.e., a greenhouse gas which is unhealthy to living organisms. Though there is a new crusade under the aegis of sustainable development goals for the promotion of clean environment through the adoption of renewable energy sources, the use of fossil fuel is still on the increase due to many factors such international politics, governmental inadequacies, corruption, and existing employments relating to fossil fuel (Fig. 2.1).

Combustion of fossil fuels is considered a major source of the increased CO_2. The amount of CO_2 produced per equivalent energy unit varies depending on the fuel-gas produces less than oil, and oil produces less than coal (Fig. 2.2). There are other sources of CO_2 production aside from fossil fuel as presented in Fig. 2.2.

Aside from the air pollution from fossil fuel, the pollutants in fuel include mercury, arsenic, and sulfur in coal; sulfur, vanadium, and nickel in oil; and sulfur in gas. These pollutants in the form of heavy metals are an extended danger of fossil fuel burning.

Wildfire: Climate change is causing an increase in forest wildfires. These wildfires have a high contribution in pollution. Wildfires could also be caused by burning of farm stubble. When these fires are ignited, they cause smog and these smog could lead to difficulty in breathing (Fig. 2.3).

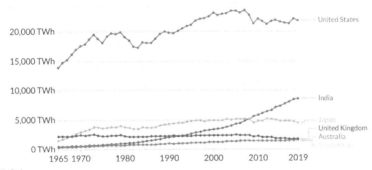

FIGURE 2.1

Fossil fuel consumption (Ritchie and Roser, 2017).

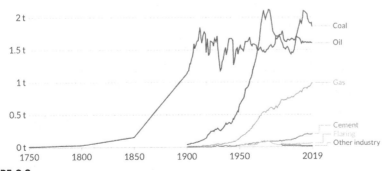

FIGURE 2.2

Carbon dioxide emission (Ritchie and Roser, 2017).

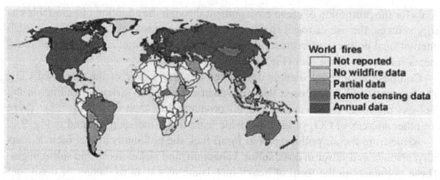

FIGURE 2.3

Global wild fire measurement.

Transportation: Transportation is another contributor to the high rate of air pollution. Due to advancement in transportation technology, the demand for faster and more efficient means of transportation increased among people. There is therefore an increase in need of fuels to facilitate transportation, which increases combustion of fuels. The combustion of these fuels lead to the excessive release of carbon (IV) oxides and other harmful gases into the atmosphere. In some cases, such as in spacecraft, radioactive particulates have been dispersed into the atmosphere through the adoption of radionuclide compounds as fuels (Ref). Also, pollution from ships off-shore have been seen to be a source of pollution that is leading to climatic variations over the years (Ref). The equation that expresses the movement of pollutants released from a source is mathematically written as (Albani et al., 2015):

$$\frac{\partial c_P}{\partial t} + \frac{\partial (u_x c_P)}{\partial x} + \frac{\partial (u_y c_P)}{\partial y} + \frac{\partial (u_z c_P)}{\partial z} = \frac{\partial}{\partial x}\left(K_x \frac{\partial c_P}{\partial x}\right) + \frac{\partial}{\partial y}\left(K_y \frac{\partial c_P}{\partial y}\right) + \frac{\partial}{\partial z}\left(K_z \frac{\partial c_P}{\partial z}\right) + R_c$$
$$+ Q_p$$

where c_p is the pollutant concentration in ($\mu g/m^3$); u_x, u_y, and u_z are the wind velocities in the different space directions, x, y, and z, respectively; K_x, K_y, and K_z are the diffusivity parameters in the different directions of space; R_c ($\mu g/m^3 s$) is the term of reaction or decay due to physicochemical reactions in the atmosphere that depend on each pollutant; and Q_p ($\mu g/m^3 s$) is the term that implements the source or sink of the pollutant. The equation was formulated to address turbulence, wind dragging, buoyancy forces, and diffusive effects. Emetere (2016) gave a model that includes turbulence, wind dragging, differential deposition, buoyancy forces, and diffusive effects:

$$\frac{\partial C}{\partial t} + V_x \frac{\partial C}{\partial x} - V_z \frac{\partial C}{\partial z} - V_y \frac{\partial C}{\partial y} = \frac{\partial}{\partial z}\left(K_z \frac{\partial C}{\partial z}\right) + \frac{\partial}{\partial y}\left(K_y \frac{\partial C}{\partial y}\right) + \frac{\partial}{\partial z}\left(K_{z2} \frac{\partial C}{\partial z}\right) + \frac{\partial}{\partial y}\left(K_{y2} \frac{\partial C}{\partial y}\right)$$
$$- P + S$$

where V is the wind velocity (m/s), P is the air upthrust, C (x,y,z) is the mean concentration of diffusing pollutants of diffusing substance at a point (x,y,z) (kg/m^3), K_y,

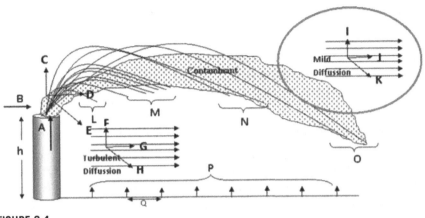

FIGURE 2.4

Model formulation diagram.

K_x are the eddy diffusivities in the direction of the y- and z-axes [m²/s], and S is the source/sink term [kg/m-s].

The diagram of the formulation is presented in the Fig. 2.4.

Burning of garbage: The open burning of garbage is also another source of air pollution. When garbage is burnt in the open, it leads to the release of some very harmful gases, and these gases when inhaled could cause harm to the health of infant, pregnant women, and the general public.

1.1.1 Causes of air pollution

- Fuel (petroleum, diesel) powered vehicles
- Burning of trees and plants
- Burning (refining) fossil fuels
- Agricultural processes-ammonia is largely emitted during such practices
- Mining
- Natural events such as volcanic eruptions, dust storms and wild fires

1.1.2 Effects of air pollution

- It leads to acid rain. This is caused by emissions of NOx and SO₂.
- It enhances eutrophication. This is a process whereby the high concentrations of nutrient are deposited into water bodies from the gases like nitrates and phosphates, which then causes them rapid growth of algae which in turn depletes the oxygen levels in the water, thereby killing its inhabitants.
- It causes depletion to the ozone layer. Gases like nitrous oxides and volatile organic compounds are released into the air and then a catalyzed by heat and sunlight which then harms the ozone layer.
- Greenhouse effect: This is a process by the heat radiation is trapped on the surface of the earth as a result of greenhouse gases.

- Smog: It is caused by intense air pollution where the air appears as a thick smoky fog with bad odor.
- Climate change and global warming.
- It increases the rate of skin diseases
- It increases the risk of skin cancer.
- It can lead to respiratory illnesses.
- It lead to the infant and maternal mortality.

1.1.3 Control of air pollution

- Reduce use of soot emitting vehicles. Encourage using other means of transportation such as walking and cycling for short distance journeys.
- Minimize the use of open fire and its products.
- Alternate fuel powered generators for clean energy sources such as solar and wind energy.
- Afforestation—planting of trees in areas of high pollution is advantageous.

1.1.4 Vehicular pollution

Vehicular pollution is significantly when ozone harming substances and other destructive contaminations are discharged into the environment from vehicles. These unsafe toxins negatively affect both the ecosystem and the human body. Constituents of vehicular pollution includes carbon monoxide, oxides of nitrogen, particles of soot, metal and pollen, sulfur dioxide other dangerous pollutants.

From Fig. 2.5, you will see a few gases leaving the fumes of the vehicles. This is the thing that vehicular pollution is majorly about. Vehicular pollution is one of the significant reasons for air pollution in light of the fact that these gases leaving the vehicles go up into the air and thus pollute the air. Vehicular pollution is one of

FIGURE 2.5

Vehicular pollution.

the most widely recognized types of pollution across the world since vehicles are fundamentally all over the place. Indeed, vehicular pollution is disregarded in light of the fact that it is from automobile. Each country experiences vehicular pollution. Iran is adjudged to have high urban contamination which includes vehicular pollution, and dust storms which is well known locally as "the 120-day wind." Schools and different offices are a few times compelled to close down during the period while the specialists disseminate face masks to residents for wellbeing. In Nigeria, there have been deaths recorded due to indoor fossil-fuel generator emission. Gases released (and inhaled) in the rush hour traffic jammed at Bombay-India is reported as smoking 51 cigarettes every day. In other words, research in vehicular pollution does not only require measurement but a well designed numerical analysis.

1.1.5 Gas flaring

Gas flares known as the burning of gas are created through different stages of oil and gas exploration. It is a main source of concern in oil producing countries as it releases significant amount of greenhouse gases. There are research works on how to convert this process for energy generation (ref). However, in a few developing countries, these gases are burned in air, thereby polluting the atmosphere and increasing the temperature of the geographical location.

Gas flaring is also defined as hydrocarbon harvesting and the procedure of combusting gas from wells. In recent times, it is regarded as a major environmental issue, contributing to approximately 150 billion meter cube.

There are three types of flaring: emergency, process, and product flaring. Emergency flaring occurs during compression failure from valve breakage. Process flaring occurs during petrochemical processes, and product flaring occurs during exploration.

There are different causes of gas flaring:

i. Natural gas carried to the surface but cannot be used as it is burned as a means of disposal
ii. Result of oil extraction
iii. Inadequate structure to put gas for industrialization
iv. Excess gas and oils after extraction
v. To avoid explosions caused by simply bottling up huge quantities of gases, flaring is used.

The effects of gas flaring includes acid rain, air pollution, influencing climate change, and reduced agricultural practice. Sulfur dioxide and nitrogen oxide emissions are the main factors of acid rain, which also are combined with atmospheric moisture to produce sulfuric acid and nitric acid, respectively. Acidification of lakes, ponds, and rivers affects both the aquatic and terrestrial organisms. Acid rain also quickens the deterioration of construction materials and paints. Flaring of gas results in the release of impurities, toxic substances that are harmful to humans. CO_2 is produced when gas is not completely burned, and it the most toxic substance to human health. Environmental implications of this gas flaring are severe because it is such an inefficient and poor use of potential fuel that pollutes air. The effects of gas flaring on climate change are significant as it is also a form of fossil fuels burning. The main component of gas flaring is carbon dioxide. By emitting CO_2, the major greenhouse gas, gas flaring contributes to global warming. The second major gas which contributes to greenhouse effect is methane, which is released when gas is vented without being burned. Gas flaring has been seen to affect agriculture as its pollutants are released into the atmosphere like nitrogen, carbon, sulfur oxides, particulate matter, and hydrogen sulfide. These pollutants deplete soil nutrients by acidifying the soil. Given the immense heat generated as well as the pH acid characteristics of the soil, there would not be any vegetation in the areas of gas flares. Temperature changes have a different effect on crops, including stunted growth, scotched plants, and withered young crops. Gas flaring has also negatively impacted upon human health due to the inhalation of toxic gases which are emitted during unfinished gas flare combustion. These gases have been connected to negative health challenges including cancer, neurological problems, reproductive issues, developmental disorders, children's abnormalities, lung damage, and skin issues. As seen from the above, there are lots of gray area in numerically modeling to nowcast or forecast gas flares (Fig. 2.6).

1.1.6 Bioaerosol production as a source of air pollution

Bioaerosols are also known as biological aerosols. They are secondary divisions of particles that are gotten from land habitats and aquatic ecosystems into the climate. It involves living and nonliving things which include organisms, distributive methods of organisms, and their waste products. They are also be said to be minute

FIGURE 2.6

Gas flaring (ESRI, 2015).

particles that can be suspended in the air and are transferred or carried around by the wind to different places of the earth. This then implies that bioaerosol either comes from organisms that have life such as animals (i.e., pets) and trees or that are the organisms themselves (e.g., bacteria). Bioaerosol also consists of dead cells. Bioaerosol causes a lot of health effects. This means that, in general, they are harmful to the health of living beings as they come from things that stay around human life and can also be spread out through the air.

There are about four types of bioaerosols:

- Pollen
- Bacteria
- Virus
- Fungus

They are all harmful to health. Some of the types of bioaerosols are living and some are nonliving.

Bioaerosols can be ejected into the earth's surface through clouds, dust plumes, and general distribution.

It was discovered that bioaerosol concentration or focus is high at confined spaces or places both indoor and outdoor. It is relatively higher in indoor environments than outdoor environments as outdoors usually have at least a little free space, unlike an indoor place where if it is enclosed there won't be any form of breathing space.

Bioaerosol has its harmful effects on the lives of living things and they are listed as follows:

- Respiratory effects (lung infections)
- Transmittable or transferrable disease
- Cancer

Cancer is a disease caused by exposure to a toxic atmosphere such as being present in a place containing bioaerosol and inhaling oxygen which has already been infected by bioaerosol. Cancer can also be gotten when people are working in poultry industries. Animals have this bioaerosol and so when people work around them and do not protect themselves, they can get the bioaerosol from the animal urines, excreta, etc. which then leads to cancer (especially cancer of the skin).

Bioaerosols cause transmittable or transferrable diseases. Here, the diseases are transferred by bacteria or microorganisms that carry diseases but we are dealing with the ones that carry the bioaerosol. This type is said to be communicable because they are gotten by direct contact such as touching and licking, and they can also be gotten via indirect contact such as through coughing.

1.2 Land pollution

This is the deterioration and eradication of the earth's surface and top layer of its soil as a result of anthropogenic activities (human activities on the earth). Land pollution causes major harm to the ecosystem. It is important to create awareness about the dangers of land pollution as it affects not only the soil but animals and human beings as well. If contamination is not reduced, organic farming will not be possible in the years to come as all the nutrient would have been removed, and there will be a large increase in the number of barren plots of land.

1.2.1 Causes of land pollution

Land pollution is typically caused by:

- Human activities like mining processes and oil drilling
- Improper disposal of waste
- Deforestation
- Excessive use of pesticides, insecticides, and herbicides
- Industrialization and expansion of cities leading to increased activities on the land such as varying soil compactions
- Factory/industrial waste
- Activities from mining
- Large amounts of heavy metal and mineral oil
- Improper disposal of industrial waste
- Oil spillage
- Extensive use of pesticides and other farm chemicals

1.2.2 Effects of land pollution

- Erosion
- Chemicals effects on the plants, damage to the ecosystem
- High risk of wildfires
- Removal of top soil and nutrients from the soil
- Ground water poisoning
- On humans, it can cause defects in children such as skin diseases and breathing disorders
- Illnesses cause by eating fruits and vegetables grown in contaminated soil
- Increase in dump sites resulting in contamination or land and breeding parasite carrying animals and insects
- It can cause degradation in vegetation
- It can cause decrease in soil nutrients
- It can cause harm to plant growth and produce

1.2.3 Control measures for land pollution

- Proper disposal of human waste
- Recycle nondegradable materials
- Afforestation, planting of trees, will help the soil and will serve as cover for other plants to flourish
- Organic farming to reduce the use of pesticides and herbicides
- Draining of septic tanks periodically to avoid contamination of ground water

1.2.4 Soil contamination

Soil contamination is the occurrence of pollutant in soil above certain recommended level that cause depletion of the soil quality and nutrient. Soil contamination can also be considered as the presence of man-made chemical that can alter natural soil function. The most common chemical pollutants in some developing countries are hydrocarbons, solvents, pesticides, and lead. Soil contamination creates a significant risk to human health. Deposition of hazardous substance from local substance could deteriorate soil and groundwater quality. The implication of these to human health is direct such as ingestion and dermal absorption. Soil contamination in developing countries is commonly located in landfill areas, oil industries, nuclear plants, and military camps (Fig. 2.7).

Soil pollution is cause by various human activities, and these activities have various negative impacts on the society. There are various impacts of soil pollution:

- Reduction in quality of soil, air, and water
- Emission of harmful rays (such as nitrogen, ammonia, hydrogen sulfide)
- Respiratory illnesses
- Reduced crop yield
- Ecological imbalance
- Increased salinity
- Reduced vegetation

  Types of soil pollution includes

- Agricultural soil pollution
- Soil pollution by industrial effluents and solid wastes
- Pollution due to urban activities
- Pollution of underground soil

FIGURE 2.7

Soil contamination.

The source of soil pollution can be of two types: agricultural sources and nonagricultural sources. Some agricultural practices lead to soil pollution such as animal husbandry, and their wastes that contaminate the soil include animal waste, long use of fungicides, and herbicides. Some nonagricultural sources can be derived from urban activities caused by rapid increase in population, which leads to increase in the waste output, thereby leading to an increase in accumulated toxic concentration. Various methods may be adopted to solve soil pollution using computer models for analyzing transport and quantity of soil chemicals. In this book, various numerical analysis shall be illustrated to show how simple this process can be. There are several remediation for cleaning soil contaminants from a given soil. They include:

- Excavation: This involves the excavating of contaminated soil and taking it to a dump site. It also involves dredging of area.
- Thermal remediation: It includes the introduction of heat to make volatile chemicals to evaporate.
- Bioremediation: This involves introduction of certain microbial to nullify the effect of some organic chemicals.
- Reforestation: Planting of soil should be practiced to reduce/control soil erosion.

1.3 **Water pollution**

This is the alteration of the natural state of water bodies by adding contaminants and toxic substances caused by human activities, thereby interfering with the beneficial uses of water. Most of water pollution is caused by human activities, but it can also be caused by natural processes. Water bodies include ponds, lakes, rivers, and oceans. Water pollution has a great effect on the environment as water is used for numerous day-to-day activities. Water pollution has seven types:

- Surface water pollution: This is the most visible form of pollution. It can be seen on lakes as oil bubbles.
- Groundwater pollution: This type affects the water below the soil and contaminates the water for drinking.
- Microbial pollution: This is caused by microorganisms in water. Most of the times, these organisms are harmless, but cause serious illnesses such as typhoid, dysentery, and cholera.
- Oxygen depletion pollution: This occurs when there is a large number of microorganisms in the water, and they use all the obtainable oxygen, thereby depleting the oxygen levels in the water and so harmless aerobic and anaerobic microorganisms will not be able to survive.
- Nutrient pollution: This is when there is excess nutrients in the water causing rapid growth of algae that take up all the oxygen harming all other life forms in the immediate environment.
- Suspended matter pollution: This occurs when the pollutants do not mix with the water, creating a harmful environment for its inhabitants.
- Chemical pollution: This is a result of chemical waste from industries flowing into nearby rivers and streams.

Advancements in water research in the past seemed to depend more on the availability of reliable data from the field than on the ingenuity of the modeler. In recent times, models have shown how interesting water research can be optimized, applied, and revamped. Implementation of the model as a tool is common in the form of mathematical models (devised an advection-diffusion water quality model, finite-difference model), computational models, observation models (ecological model, hydrodynamic model), and statistical models. Among the first models in water research are Delaware Estuary Comprehensive Study model, DECS model, Streeter-Phelps model, DOSAG model. Delaware Estuary Comprehensive Study model seeks to control water pollution in a city; DECS model seeks to create deeper understanding on rates of biodegradation and reaeration. Streeter-Phelps model seeks to create in-depth understanding on dissolved oxygen (DO) concentration and biochemical oxygen demand (BOD). DOSAG model solved the steady state problem for a multisegment river system. QUAL 1 model is to simulate stream temperature as well as DO and BOD; QUAL 2 model is used to simulate more complex stream systems i.e., for both steady and unsteady flow. Both QUAL 1 and 2 are used to evaluate the impacts of nutrient loading on the oxygen resources of streams. WRE and MIT models are used to simulate reservoir as a one-dimensional system of horizontal slices to simulate the vertical distribution of heat. CLEANER and LAKECO are ecological models used to access one-dimensional temperature transition in single and multisegment systems respectively.

1.3.1 Causes of water pollution

- Chemical wastes from factories like breweries and soap manufacturing companies
- Oil spillage
- Improper waste disposal
- Eutrophication
- Domestic sewage
- Plastics
- Agricultural practice such as washing fertilizers and pesticides into river bodies
- Disposal of industrial waste

1.3.2 Effects of water pollution

- Contaminated water can cause and spread waterborne diseases like cholera and dysentery
- Poses danger to aquatic life
- Chemically contaminated water can pollute the soil causing soil infertility
- Drinking water with excess amount of fluoride can have harmful effects on the spinal cord
- Acid rain makes water bodies unconducive for aquatic animals
- Eutrophication
- Increased harm to marine life
- Increased waterborne diseases
- Disruption to the ecosystem

1.3.3 Control measures for water pollution

- Cleaning up areas around lakes and beaches to avoid plastics entering the water bodies
- Support laws against pollution
- Abstain from flushing contaminated liquids, pills, and drugs down the toilet
- Reduce the use of pesticides and fertilizers
- Reduce the use plastics

1.3.4 Well pollution and why it is still a problem in developing countries

A well is a hole that is formed by drilling or digging in order to be able to access liquid resources; it could be hydrocarbon wells or groundwater wells (that contain water) that are being sourced. Groundwater wells are the oldest and most widely drilled wells for obtaining water from the subsurface. Water obtained from wells is used mostly in rural area for domestic uses such as cooking and drinking. Well source contamination occurs when natural or man-made chemicals and products find their way into the wells and cause them to become harmful for human consumption (Fig. 2.8). When drilling for wells, steps are taken to perform the geophysical survey to find out aquifert zone, i.e., water catchment that is a saturated zone that can produce enough amount of water that can serve the purpose for which it is being drilled. This process leads to land pollution as certain radioactive substance beneath the earth surface is displaced above the earth surface. Several researchers have shown this over time (Ref). Well pollution occurs as a result of toxins being released into existing underground water bodies called aquifers. When contaminants are released into the atmosphere, they contaminate the well. The well pollution being focused on is water pollution. Well contamination sources are nitrates which are

FIGURE 2.8

Contaminated well.

sewage that have seeped into the wells due to the poor infrastructure of the wells and also due to pit latrines which may be present close to the wells. Another source is lead which can come from lead pipes which may be present in the well.

It is a form of water pollution that is primarily induced by the deliberate or unintentional release of substances because of anthropogenic or natural causes. Biological, physical, and chemical components of the pollutants is passed into aquifers. The flow is aided by processes such as diffusion, dispersion, absorption, and the speed of flowing water.

Water body pollution has several intricacies that make it very difficult to monitor. Some of the example include increase of nonpoint source pollution, uneven diffusion of air pollution in water, concentration of contaminants into water bodies at undefined times, atmospheric deposition etc. Scientists have contributed in solving different challenges. Delong and Huiping (2018) investigated many water quality models to determine the structure and parameters of the water quality model i.e., atmospheric pollution source, water quality model, water pollution source, water environment, and water classification. The researcher used the differential equation of the concentration change of a pollutant in the water injection fluid

$$\frac{\partial C}{\partial t} + \mu \frac{\partial C}{\partial x} = E \frac{\partial^2 C}{\partial x^2} - kC + R \tag{2.1}$$

where C is water concentration of a certain pollutant in mg/L; t is time in d; μ is the average flow rate in m/d; x is the water body distance in m; E is the dispersion coefficient of flowing water in m^2/d; k is a pollutant attenuation rate coefficient in d^{-1}; R is a system of internal factors in mg/(L.d).

Busayamas et al. (2009) worked on simple mathematical model that incorporated the reaction—diffusion—advection equations for the pollutant and dissolved oxygen concentrations to understand the effect of aeration on the degradation of pollutant for river pollution. The researcher coupled equations (such as zero dispersion, dispersion in linear kinetics) to expand the rates of change of the concentration with position x and time t expressed as

$$\frac{\partial(AC)}{\partial t} = D_p \frac{\partial^2(AC)}{\partial x^2} - \frac{\partial(\nu AC)}{\partial x} - K_1 \frac{X}{X+k} AC + qH(x) \tag{2.2}$$

The coupling of these equations became possible because of the reactions between oxygen and pollutant to produce harmless compounds. Based on the above, the effect of aeration on the degradation of pollutant was solved using modified models.

It is observed that the modified equation of Eq. (2.2) was found to have direct application in understanding eutrophication. The equation was documented in Balcerzak and Zimoch (1997) as

$$\frac{\partial(AP)}{\partial t} = \frac{\partial}{\partial x} \left(-U_X AC + E_x A \frac{\partial C}{\partial x \partial t} \right) + A(S_l + S_b) + AS_k \tag{2.3}$$

where U_X is the longitudinal speed of advection [m/d], E_x is the coefficient of dispersion in the direction of water flow [m²/d], S_l is the direct and diffusive external contamination load [g/m²d], S_b is the load exchanged between segments as a result of longitudinal dispersion [g³/md], S_k is the contamination load due to kinetic transformations [g/m²d], A is the transformation area [m²].

Eq. (2.3) allows the determination of the growth, death, and control of algae during eutrophication. Kachiashvili et al. (2007) worked on the transport of pollutants in rivers using one-, two-, and three-dimensional models in terms of time-dependent convection—diffusion—reaction differential equations as presented in equation 2.4. The model was verified understudy the diffusion and transport of chemicals within water bodies. It was reported that the successes of the model was to determine concentration despite the geometry of the rivers, the arrangement of the control sections, and the concentrations of polluting. Hence, using theories, more challenges on river pollution dispersion can be resolved.

$$\rho^i \partial \psi^i / \partial t + \mathbf{u} \cdot \nabla \psi^i - \nabla \cdot \left(k^i \cdot \nabla \psi^i \right) + R^i = 0 \tag{2.4}$$

where (x, t) is the ith dependent (unknown) variable. The velocity field is represented by u, and ψ and κ are the accumulation capacity and diffusion tensor for the *i*th variable. The source/sink term R' in our cases is due to chemical reaction.

Pollutants travel slowly through an aquifer; their concentrations appear to be high which results to a plume.

Well pollution is therefore caused by

1. Natural sources
2. Septic systems
3. Hazardous waste disposal
4. Petroleum products
5. Injection wells

Contamination can occur when naturally occurring compounds in soils and rocks dissolve in water. Some of those compounds are sulfates, iron, radionuclides, fluorides, manganese, chlorides, and arsenic, but others such as decaying soil materials, can seep into the underground water and travel as particles with it. Septic systems are the leading source of groundwater contamination all over the world. Poop, septic tanks, and cesspools all add to the waste. Because of the large number of people who depend on the scheme, it is one of the most polluting sources. In addition nitrates, oils, bacteria, chemicals, and so on are released into underground water by poorly built and leaking septic systems. Injection wells can be used for several purposes but when not used properly or managed properly, dangerous substances can be put into the wells from the injection well. Due to this fact, if the injection wells are not managed properly then it can lead to contamination and pollution of the well.

1.3.4.1 Dangers of well pollution
The dangers of well pollution refers to the things that make well pollution dangerous. Some of the dangers are

1. Risk of acute and chronic toxicity
2. Liver damage
3. Kidney damage
4. Anemia
5. Cancer

1.3.4.2 Policy of well pollution

Pollution prevention is reducing or eliminating waste at the source by modifying production processes, promoting the use of nontoxic or less toxic substances, implementing conservation techniques, and reusing materials rather than putting them into the waste stream.

Pollution prevention means source reduction and EPA defines P2 in this Memorandum—May 28, 1992, Subject: EPA Definition of "Pollution Prevention."

Well pollution is still a problem in developing countries because of lack of adequate resources and cost of implementing the necessary procedures to ensuring well pollution is reduced or stopped entirely.

1.3.5 Bore-hole pollution

Borehole is one of the capital or most essential sources of water, and it is found underneath the surface of the earth. It is most exposed to pollution or contamination, and this contamination can be due to man-made products such as gasoline, oil, and chemicals, when these chemicals diffuse into the ground. They become not healthy for human use, and these are problems that affect developing countries. The main sources of borehole contamination and diffusion are in farming chemicals, storage tanks, and atmospheric pollutants. The most necessary natural resource around the globe or world is the ground water. According to research studies, it makes 30% of the world's freshwater reserve. Why is borehole an important topic for discussion? (Fig. 2.9)

- Borehole is clean most of the times, it is easily accessible, and it is the most important source of drinking in the world.
- Borehole is needed by farmers in the farm for areas where there is irrigation for crops.
- Countries that have dry regions like Australia, borehole provides a cheap water source because it is the most price-effective for the countries.

Sources of borehole contamination includes:

1. Septic system
   This is a chamber made of concrete or plastic that is dug under the ground through which sewage flows for treatment. All waste water disposal systems are supposed to be connected to the city or town sewer system, but not all of them are connected. The ones that are not connected have the tendency to leak some unhealthy organisms or chemicals to the borehole, and this causes health problems to humans.

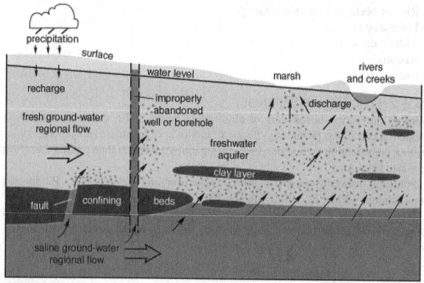

FIGURE 2.9

Borehole contamination.

2. Landfills

   A landfill is a place where garbage or waste are buried. Landfills are supposed or expected to have a boundary between the waste material and the subsurface to avoid contamination getting into the water, but if there is no boundary, the contaminants from the land fill will get into the water.

3. Storage tanks

   Storage tanks are where oils and gasolines are stored and are either above the ground or below the ground. It has been estimated that around the globe, millions of tanks are buried, and this tanks are left overtime and the metals rust and become weak. It develops leaks, and this oil spills into the ground and gets to the water and serious contamination can occur.

   The movement of water between subsurface can act as dependable natural blockage to contamination but only works under beneficial conditions.

   In any area, the stratigraphy of that area plays a vital role in the movement of pollutants. Areas of topography from the dissolution of soluble rocks on limestone bedrock can become prone to subsurface contamination from boreholes. Earthquake faults can also be passages through which contaminants go downwards and the water table is another medium through which contaminants can come into the ground because they are of good essentials for drinking water and waste disposal. Nuclear waste and many of these chemicals go through chemical change or decay, and this lasts for a long period of time in the water reservoirs under the ground.

Prevention of borehole contamination sources can be achieved from the following:

- Land zoning for boreholes protection: The improvement at which lands are separated and shared into zones has helped a lot in protecting our boreholes from contaminants, and this has started taking place is most countries by water authorities.
- Source protection map: This makes reference to the capturing of areas around an individual borehole source to protect it from pollution. Thus, the potential sources of pollutants can be noticed at a distance which moves around that same flow path are long enough for the pollutant to be eliminated through filtration.
- Quality monitoring of borehole: All this monitoring has been executed recently in many countries, and it is used for the development of conceptual models.

1.3.6 River contamination

River contamination is said to be polluting the water bodies such as rivers normally due to human activities. Humans dump domestic waste, bicycles, garden cuttings, and electronic waste into rivers or river banks (Fig. 2.10). Aside from pollution of the water ways, river contamination also harms wildlife, aquatic life, and increases the risk of flooding. River contamination is a source of concern worldwide, especially in developing nations where about 3.2 million die as a result of contaminated drinking water and poor sanitation. Globally, about 829,000 people die every year from diarrhea caused by river contamination. WHO also reported that 2.5 million death cases occur due to dysentery to which river contamination is a significant source. River contamination is caused by several factors not limited to sewage, radioactive substances, agricultural pollution, oil pollution, etc. River contamination could lead to diseases such as cholera and typhoid; lead to the demolition of

FIGURE 2.10

River contamination.

ecosystems; lead to disorderliness in the food chain; lead to the death of aquatic organisms and sometimes humans. It allows eutrophication to take place.

Sewage are liquid refuse or waste matter that is usually carried off by sewers. It may be the debris from industrial and indigenous (domestic) processes. From research, it was found that not minding the strict regulatory control, researchers have shown that the water and sewage industry accounted for almost a quarter of the serious water incident in England and Wales in 2006. Radioactive substances are substances that are found in the ground because of the radioactive elements commonly found in rock and different types of soil. They are used in industrial, medical, nuclear power plants, and other scientific processes. They are seen in televisions, luculent clocks, wristwatches, and X-ray machines. Its effect of its pollution can last a lifetime as it causes cancer, impaired growth in babies. Agricultural pollution relates to the derivatives from farming practices caused by living and nonliving organisms that lead to polluting of the environment and ecosystems around which causes damage to humans and their gainful interests. Examples are unguided distribution of slurries and manure, tillage, pesticide leaching, fertilizers, and also spills from milk diaries. Oil spillage is the flow of a liquid petroleum hydrocarbon into the environment, most especially the water systems such as rivers, which is usually because of human activities. This oil spillage makes drinking water unhealthy to drink, and it causes death to wildlife and aquatic organisms. It decreases the oxygen level within the aquatic environment. Oil spilling is caused by the following factors:

- Accidents involving oil tankers, pipelines, refineries, etc.
- Carelessness from people handling oil production, for instance when there is leakage during delivery of oil
- Breaking down pieces of equipment
- Natural disasters, i.e., hurricanes
- Deliberate acts by terrorists, illegal dumpers, etc.

1.4 Noise pollution

Noise pollution is the release of excessively loud noises to the environment. Any sound above 85 decibels can be said to be dangerous to us as humans.

1.4.1 Sources of noise pollution

- Industrial noise from machines
- Transportation, i.e., noise from vehicles
- Household noise, i.e., noise from household appliances
- Noise from other loud devices such as speakers and, any more

1.4.2 Effects of noise pollution

- It can lead to deafness
- It can lead to reduced hearing
- It could lead to problems in communication

1.5 Radioactive pollution

This can be defined as pollution that is caused as a result of the release of radioactive materials into the environment. Its sources could be from nuclear power facilities, military establishments, research organizations, hospitals, and general industry. In addition, historical tests of nuclear weapons in the atmosphere and underground, nuclear and radioactive accidents, and the deliberate discharge of radioactive wastes from nuclear and other installations represent sources for radioactive pollution. Such radionuclides have the potential to find their way from air and water onto the ground and into the food chain. Radioactive pollution most of the times refers to the exposure of human to ionizing radiation (USNAS & NRC, 2006). The U.S. Environmental Protection Agency (USEPA) clearly reported that the average human receives about 620 millirem (mrem) of ionizing radiation per year (USEPA, 2012), which may likely be from natural influx from space, rocks, soils, synchrotron/accelerator facility, nuclear power plant, and health care procedure. Rocks have low-level of natural radioactivity (see Table 2.1), and soils have varying low and medium levels of natural radioactivity. The most lethal state of radionuclides in soil is the release of radon in gas phase. It is reported that this type of pollution has high mobility and can accumulate in confined space. The typical annual dose from food and drinking water amounts to approximately 30 mrem (USNRC, 2017).

Another source of radioactive pollution is from radioactive waste and contaminated areas with naturally occurring radioactive material (NORM) and technically enhanced NORM (TE-NORM). Table 2.2 shows the sources of NORM.

Table 2.1 Radionuclides in rock in Florida (Missimer et al., 2019).

Isotope	Half-life	Location found	Activity
^{231}pa	3.276×10^4 years	Phosphate rock, Central FL	0.5–10.9 pCi/g
^{210}pb	22.3 years	Phosphate rock, Central FL	11.9–161.4 pCi/g
^{210}pb	22.3 years	Unfiltered groundwater, Central FL	0–7.70 pCi/L
^{210}po	138.376 days	Phosphate rock, Central FL	11.0–134.5 pCi/g
^{210}po	138.376 days	Unfiltered groundwater, Central FL	0–2570 pCi/L
^{226}Ra	1600 years	Phosphate rock, Central FL	13.3–165.2 pCi/g
^{226}Ra	1600 years	Unfiltered groundwater, Central FL	0–9.5 pCi/L
^{222}Rn	3.8235 days	Unfiltered groundwater, Central FL	6730–19,000 pCi/L
^{230}Th	7.538×10^4 years	Phosphate rock, Central FL	13.6–198.8 pCi/g
^{232}Th	1.4×10^{10} years	Phosphate rock, Central FL	0.2–2.2 pCi/g
^{234}U	24.1 days	Phosphate rock, Central FL	13.3–250.8 pCi/g
^{238}U	4.5×10^9 years	Phosphate rock, Central FL	12.8–252.5 pCi/g

Table 2.2 Sources of NORM contamination.

Mineral ores and extracted materials		Other processing/manufacturing
Copper	Titanium	Water treatment
Aluminum (bauxite)	Tungsten	Sewage treatment
Fluorspar	Vanadium	Spas
Gypsum	Zircon	Paper and pulp
Iron	Coal (and coal ash)	Ceramics manufacture
Molybdenum	Oil and gas	Paint and pigment manufacture
Phosphate	Geothermal energy	Metal foundry
Phosphorous	Uranium and thorium	Optics
Potassium		Incandescent gas mantles
Precious		Refractory and abrasive sands
Rare earth		Electronics manufactures
Tin		Building materials

Radon emissions have been reported to be enhanced by the ambient environmental conditions such as increasing moisture contents, temperature, and specific surface area (Sakoda et al., 2011). Also, naturally occurring radionuclides such as radon can be found in public water and its effect can be very dangerous if not treated. It could also be found in groundwater and some building materials. WHO reported the reference level for airborne indoor radon to be 100 Bq/m3 or 2.7 pCi/L when converted (WHO, 2011). On the other hand, USEPA do not have reference level for groundwater or indoor air quality but believe that when indoor air quality exceeds 4 pCi/L, there could be grave implications to the human health. The transformation of radon at different scenario is presented in Table 2.3.

Radioactive pollution is specific a challenge in different parts of the globe. For example, in Lee County, uranium and radium in groundwater is given as 119 and 50 pCi/L, respectively (Kaufmann and Biss, 1977). This development could have accounted for cases of cancer patient and other ailments in the area (USHHS, 2005; ALA, 2016). The alpha emissions associated with uranium at activities above 50 pCi/L are basically assumed to be very high (Roessler et al., 1979) because uranium concentrations exist in an 234U/238U activity ratio of equal or less than 1 (Levine, 1988).

1.5.1 Causes of radioactive pollution
- Nuclear explosions
- Improper disposal of radioactive waste
- Mining of radioactive ores

1.5.2 Effect of radioactive pollution
- Deaths

Table 2.3 Radon transformation in different medium (Missimer et al., 2019).

Aquifer	Activity range for radium-226 (in pCi/L)	No. of samples	Percentage of total
Surficial	Not detected	0	0
	0–5	6	75
	5–10	2	25
	10–20	0	0
	>20	0	
Intermediate-zone 1	Not detected	1	6.25
	0–5	7	43.25
	5–10	6	37.5
	10–20	1	6.25
	>20	1	6.25
Intermediate-zone 2	Not detected	0	0
	0–5	8	24.3
	5–10	11	33.3
	10–20	10	30.3
	>20	4	12.1
Intermediate-zone 3	Not detected	1	10
	0–5	5	50
	5–10	3	30
	10–20	1	10
	>20	0	0
Intermediate-zone 4, floridian aquifer systems—zones 4 and 5	Not detected	0	0
	0–5	4	57.1
	5–10	3	42.9
	10–20	0	0
	>20	0	0

1.6 Electronic waste pollution

There are different types of waste which include municipal wastes, hazardous wastes, and biomedical wastes. Municipal wastes include household and commercial wastes while hazardous wastes are usually industrial wastes.

Electronic waste includes wastes from mobile phones, laptops, computers, and television. Due to the technicality of these products, most of them are preferred to be replaced instead of being fixed when damaged. Batteries are a main electronic waste. Apart from being able to recharge batteries, they contain harmful sulfuric acid that is hazardous when left out in the atmosphere like that.

Biodegradable materials are basically materials that can break down or be decomposed by bacteria or microbes by natural factors such as temperature, oxygen, and ultraviolet. Most of these materials are kitchen or food wastes. They break down into simpler constituents and eventually fade into the soil. Other examples include green waste, municipal waste, manure, etc. Unlike biodegradable materials, these cannot be easily broken down. They cannot be broken down into simpler constituents. Most of them take thousands of years to degrade. Hence, this leads to vast pollution due to their cheapness and ready availability. Due to this pollution, they contribute in causing cancer and other health challenges to both humans and animals. Most inorganic materials are not biodegradable and instead they can be recycled.

The importance of health to man and animal can never be over emphasized. None is able to function properly in ill-heath. Whatever is therefore capable of affecting human health should be addressed adequately.

Refuse emerges from improper disposal of unwanted materials. This improper disposal often leads to pollution when done on a larger scale. Heaps of these unwanted dumped materials tend to help in the rearing of rodents, snakes, insects, etc. When these materials are dumped closer to the water bodies, they tend to pollute and soil the water ways making it unclean for human consumption. It also pollutes the natural habitats of fishes. This stagnant water that is being produced as a result of the introduction of materials in it tends to support the development of waterborne diseases which include dysentery, cholera, etc. Apart from this, it also results in air pollution; stagnant water causes odor and breeding of insects such as mosquito.

Pollution also uses up space meant to do more productive things. This may even cause road blockage which creates more work for government when trying to clear out the roads.

Electronic waste also known as e-waste is basically waste from electronic products or materials. Due to the versatility of these materials, they are preferred to replace instead of being fixed for the fear of glitches. Some of these wastes are hazardous. A big example of this is the cathode ray tubes from the televisions and monitors. Below are a group of items with their hazardous components (Table 2.4).

Table 2.4 Contamination from e-waste.

Item	Hazardous components
Cathode ray tube	Lead, antimony, mercury
Liquid crystal display	Mercury
Circuit board	Beryllium, antimony
Fluorescent tube	Phosphorous
Cooling system	Ozone depleting substance
Plastic	BFR
Insulation	Asbestos, ozone depleting substance in foam
Rubber	Lead, lithium
Batteries	Phthalate, BFR lead

Mercury causes a form of impairment in neurological development. This may be in the fetus in children. Other problems include insomnia, headache, and cognition. Sources of these are switches, fluorescent, and monitors. Lead causes a mild damage in the brain and in the nervous system. It could also cause child bearing issues in women. Sources of these include circuit boards and monitors. Chromium may cause perforated eardrums, kidney damage, ulcer, and sometimes discoloration of the teeth and pulmonary congestion. Sources of these include untreated and galvanized steel. BFR is a major cause of cancer to the digestive and lymph systems. This could also cause endocrine disorder. Sources of these include plastics and circuit boards. Cadmium is usually released in gases. If this gas is inhaled; it could cause great damage to the lungs, kidney, and the mental cognition. Sources of these are light sensitive resistors and Ni–Cd batteries. Pollution management are usually groups or set of organizations that come together strictly to combat the emission of pollutants (materials) into the environment. These are usually very identifiable. Due to the current environmental pollution, some strategies have been put in place by government, individuals, and communities to control the hazard. However, we are also told to aid in less pollution by doing the following (Table 2.5):

- Proper use of equipments: When devices are well taken care of the tendencies of them getting spoilt becomes very low.
- Proper checking of equipments: When devices start malfunctioning, they should be taken to professionals for maintenance check. This is to ensure that they are fixed properly and not further damaged when taken to an amateur.
- Reuse of fixed devices.

Table 2.5 International agencies and their functions to help combat pollution.

Organizational agencies	Functions
The Basel Convention	- A mobile phone partnership - E-waste global initiative - E-waste partnership on computing equipments
G and 3Rs	- Reduce, reuse, recycle
StEP (Solving the E-waste Problems)	- Offspring of United Nations - To optimize life cycle of EEE - To reduce environmental risk
UNEP/DTIE (IETC)	- Implementation of solid waste management - Reduce, reuse, recycle of all forms of waste product - Integrated solid waste
GeSI (Global e-Sustainability Initiative)	- Objective to share experience and knowledge - Consists of ICT service providers
GTZ	- E-waste management to all countries - Indo-European e-waste initiative

- Recycle: This should be done when it is very noticeable that the device cannot be used again. There are often properly checked with professionals to be sure that they can no more be used. Sometimes the components of the full device can be reused while the rest will be recycled to help make another electrical device.

More awareness on how to examine environmental measurements is essential to understand the effects of pollution on our environment as it is one of the leading causes of death.

References

American Lung Association (ALA), November 3, 2016. Lung Cancer Fact Sheet. Reviewed and Approved by the American Lung Association Scientific and Medical Editorial Review Panel. Available online: https://www.lung.org/lung-health-and-diseases/lung-disease-lookup/lung-cancer/resource-library/lung-cancer-fact-sheet.html. (Accessed 28 May 2021).

Albani, S., et al., 2015. Twelve thousand years of dust: The Holocene global dust cycle constrained by natural archives. Climate Past 11, 869–903. https://doi.org/10.5194/cp-11-869-201.

Balcerzak, W., Zimoch, I., 1997. Mathematical modelling of water quality variations. Ochrona ≈örodowiska 3, 69–72.

Busayamas, P., Sweatman, W.L., Wake, G.C., Triampo, W., Parshotam, A., 2009. A mathematical model for pollution in a river and its remediation by aeration. Appl. Math. Lett. 22, 304–308.

Emetere, M.E., 2016. Numerical modelling of West Africa regional scale aerosol dispersion. A doctoral thesis submitted to Covenant University, Nigeria, pp. 65–289.

ESRI, 2015. ArcGIS 10.5: Using ArcGIS Spatial Analyst. Software User Guide ESRI, USA.

Delong, W., Huiping, Z., 2018. Water environment mathematical model mathematical algorithm. IOP Conf. Ser. Earth Environ. Sci. 170, 032133.

Kachiashvili, K., Gordeziani, D., Lazarov, R., Melikdzhanian, D., 2007. Modeling and simulation of pollutants transport in rivers. Appl. Math. Model. 31, 1371–1396.

Kaufmann, R.F., Biss, J.D., 1977. Effect of Phosphate Mineralization and the Phosphate Industry on Ra-226 in Groundwater of Central Florida. EPA/520-6-77-010. U.S. Environmental Protection Agency, Washington, DC, USA.

Levine, B.R., 1988. Uranium Isotope Survey of the Groundwater System in Lee County, Florida. Master's Thesis. Department of Geology, Florida State University, Tallahassee, FL, USA.

Missimer, T.M., Teaf, C., Maliva, R.G., Danley-Thomson, A., Douglas, C., Hegy, M., 2019. Natural radiation in the rocks, soils, and groundwater of Southern Florida with a discussion on potential health impacts. Int. J. Environ. Res. Publ. Health 16, 1793.

Ritchie, H., Roser, M., 2017. Fossil Fuels. Published online at OurWorldInData.org. Retrieved from: https://ourworldindata.org/fossil-fuels.

Roessler, C.E., Smith, Z.A., Bolch, W.E., Prince, R.J., 1979. Uranium and radium in Florida phosphate material. Health Phys. 37, 269–277.

Sakoda, A., Ishimori, Y., Yamaoka, K., 2011. A comprehensive review of radon emanation measurements for mineral, rock, soil, mill tailing and fly ash. Appl. Radiat. Isot. 69, 1422−1435.

U.S. Department of Health and Human Services (USHHS), 2005. News Release: Surgeon General Releases National Health Advisory on Radon. U.S. Department of Health and Human Services, HHS Press Office, Washington, DC, USA.

USEPA, 2012. U.S. Environmental Protection Agency. Radiation: Facts, Risks, and Realities. U.S. Environmental Protection Agency, Washington, DC, USA.

USNAS & NRC, 2006. U.S. National Academy of Sciences; National Research Council; Committee to Assess Health Risks from Exposure to Low Levels of Ionizing Radiation. Heath Risks to Exposure to Low Levels of Ionizing Radiation, BEIR VII Phase 2. National Academies Press, Washington, DC, USA.

U.S. Nuclear Regulatory Commission (USNRC), 2017. Doses in Our Daily Lives. USNRC, Washington, DC, USA.

World Health Organization (WHO), 2011. Guidelines for Drinking-Water Quality, fourth ed. World Health Organization, Geneva, Switzerland.

Further reading

Pellicer, T., O'Connell, E.J., 1981 Alva Radiological Survey. Lee County Health Department, Lee County, FL, USA.

U.S. Environmental Protection Agency (USEPA), 2013. Consumer's Guide to Radon Reduction. How to Fix Your Home. EPA 402/K-10/005. U.S. Environmental Protection Agency, Washington, DC, USA.

Typical environmental challenges

3

1. Introduction

In this section, various environmental conditions or parameters that have a wide extensive intricacy in environmental science were considered. Not all of the parameters that exist in environmental science. This section introduces a concept that would be examined more critically in preceding chapters on numerical analysis.

1.1 Thermal comfort as a source of environmental concern

Thermal comfort is a state of mind that is evaluated by subjective assessment and reflects satisfaction with the thermal environment. Thermal comfort can also be defined as a person's own awareness of the thermal atmosphere, and it is defined as a person's neutral feeling in relation to a given thermal environment, without sweating. In tropical region where the temperature and humidity tension are still high, human thermal discomfort is a serious concern. The person's location, as well as the climatic conditions within and outside the enclosure, influence their thermal comfort standards (Balbis-Morejón et al., 2020). The tropics experience heightened humidity owing to the temperature that causes perspiration and discomfort. Also, due to climate change, the cooling potential of natural ventilations that has been a technique of choice for mitigating the effect of thermal discomfort in hot humid regions would fall, and there would be rise in outdoor temperature. Protection against exposure to excessive sunshine can be achieved by providing natural devices such as trees, shrubs, and land form and external devices such as window blind, louvers, roof hang, heat absorbing glasses. 100% thermal comfort cannot be obtained in any climatic condition. Thermal comfort in buildings can be seen as the outcome of the well-balanced combination of building systems. Thermal comfort in a building can be caused by the location of the building and the activity performed inside the building. The design of the building and the indoor air quality can also affect the thermal comfort (Fig. 3.1).

The methods for measuring thermal comfort includes Predicted Mean Vote (PVM) method, Percentage of People Dissatisfied (PPD) method, ASHRAE Standard 55 etc. Fanger's PMV method is commonly used method for objectively evaluating thermal comfort. The PMV method was developed in the 1970 and are based on a constant

Numerical Methods in Environmental Data Analysis. https://doi.org/10.1016/B978-0-12-818971-9.00004-1

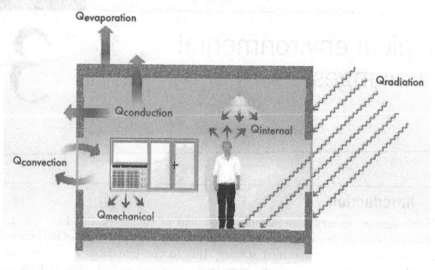

FIGURE 3.1

Diagram illustration of thermal comfort (Ecophon, 2020).

state model. When a wide number of people were used to a certain temperature, the PMV values represent their average thermal sensation. Six variables that affect a person's thermal comfort must be calculated in order to quantify PMV. Humidity, air velocity, mean radiant temperature, air temperature are four environmental variables, while occupants' metabolic rate and fabric insulation (CLO value) are two personal factors. Fanger's PPD method is the commonly used method for objectively evaluating thermal comfort. The PPD (Percentage of People Dissatisfied) method was developed in the 1970s and are based on a steady state model. ASHRAE Standard 55 necessitates the calculation of air humidity, air velocity and temperature, measuring the mean radiant temperature from three different heights at 0.7m of height 151 mm globe diameter (Limbachiya et al., 2012).

There are parameters that are used to determine the thermal comfort over an area i.e., regular variation air circulation, relative humidity, air velocity, air temperature, radiant temperature, clothing insulation, wellbeing and sicknesses, and metabolic heat or level of activity. Regular variation air circulation is related to wind system. The northeasterlies trade winds could attain a height of around 3000 m. It comes with dusty wind that originates from solid air masses from the desert area. Relative humidity is the resultant consequence of warm water evaporating into the ambient atmosphere, with the remaining water in the air becoming the humidity. Air velocity affect rapid loss of heat for the average parts of the tropical region. Evaporation is boosted, and the body is cooled. Air temperature is associated to the weather. It's marked by ups and downs. High temperatures, on the other hand, induce thermal discomfort, affecting lifeform in tropics. One of the most significant factors that influences thermal comfort is thermal radiation, that is, the heat that a warm object radiates. Where there are heat sources in an area, radiation is emitted. As a result, it has an effect on how one loses or gains heat from the atmosphere. Clothing obstructs the capacity to lose heat to the atmosphere by its own existence. The insulating effect of clothing on the wearer is essential to thermal comfort. Wellbeing and sicknesses play specific roles in thermal comfort. Colds and flu, for example, hinder our capacity to sustain a central body temperature of 36.5°C. Metabolic heat or level of activity also plays significant roles in thermal comfort. The heat that we create as a result of our physical exertion. As compared to someone who is exercising, a person who is stationary would feel cooler.

Thermal comfort can be analyzed by the following methods such as control volume and discretized analysis. The control analysis method creates a boundary around a volume and solve for the input and output. The methods listed above might not give the same output because they analyze the problem in different ways. There are software programs that are used to evaluate thermal comfort in a space. This tools/software vary in their scope, capabilities and limitation. The software can be categorized based on the following: analysis method, scale of focus, spatial output, and temporal output (Fig. 3.2).

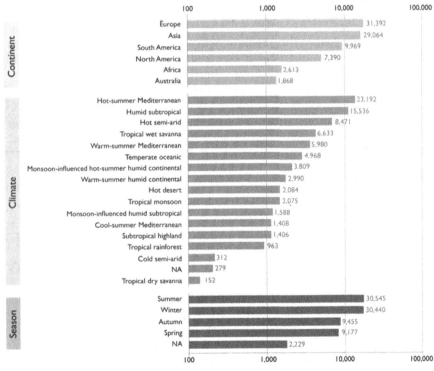

FIGURE 3.2

Thermal comforts across the continents (Ličinaa et al., 2018).

1.2 Rainfall as a source of environmental concern

Rainfall is the quantity of precipitation that falls as rain (water from clouds) on the Earth's surface, whether on land or water. When air masses pass over warm water bodies or wet land surfaces, it forms. Moisture or water vapor is carried upward into air masses by atmospheric turbulence and convection, where it form clouds. This water vapor is finally released by the clouds and falls in the rain. The challenges of rainfall system are not limited to the parameter mentioned but includes damage to buildings and infrastructure, loss of crops and livestock, flood, risk to human, life, increased erosion, landscape degradation, the quality of fresh water produce has been affected, biodiversity loss. Rain is the form of water droplets when precipitated in the atmosphere, and is induced by the water vapor condensation and obey the law of gravity because of the water weight as it falls freely. It is an important activity of the water cycle and not a negligible source of the Earth's fresh water. In fields such as agriculture, hydrology, and forestry, rain measurements and analysis play a vital role.

Rain measurements is acquired using two methods known as manual method and digital method such as automatic weather station. This is the most used and simple method of rainfall measurement. In this method, the measurement of rainfall is taken by an instrument called a rain gauge. Rain gauge operates by taking the rainfall incidence at a point through an opening of the area. There are two types of rain gauges:

a. Recording rain gauge: This produces automatically a permanent record of the rainfall and also records the cumulative rainfall. It operates in a mode such that the total amount of rainfall required is immediately recorded in a graph paper which produces the plot of cumulative rainfall against time that represents the mass curve of rainfall. It also measures the duration and intensity of rainfall.

b. Non-recording rain gauge: This comprises of a collecting area of diameter 12.7 cm to collect rainfall and is in a circular form. It is the simplest type of rain gauges and they collect rainfall but quantity of collected rainfall is not recorded.

The methods of analyzing rainfall include the following:

1. Hyetograph method: This is a bar graph that depicts the intensity of rainfall versus the time interval. It is obtained from the plot of cumulative rainfall against time which represents the mass curve of rainfall or from the data gotten from the rain gauges, it is used to predict floods and storms.

2. Frequency analysis of rainfall data: This is used to estimate the rainfall depth's magnitude from rainfall data which depicts the rainfall depth that can be expected on probability periods. It is used mostly in agriculture and hydrology. It enables good computations and conservative plotting.

3. Mass curve of rainfall method: This is the plot of the cumulative depth against time interval and is in a chronological order. The intensity and total depth of rainfall can be determined, the indication of no rainfall is depicted by a horizontal line in the curve and intensity of rainfall is depicted by steepness of the curve.

4. **Point rainfall method:** This is the interpretation of rainfall data gotten from a specific station. It is depicted graphically in a bar graph of magnitude of rainfall against time.
5. **Chemical analysis:** This is the use of the Atomic Absorption spectrometer to analyze water samples collected from rainfall.

The challenges of rainfall system include the following:

1. Reduction of food and food insecurity in communities in Nigeria
2. Large scale flooding in the southeast and southwest
3. Aridity, dryness, and desertification in the North due to the low level of rainfall
4. Poor agriculture system when the rainfall level is much lower and higher than the average level of the normal, the growing season is affected and problems occur e.g., low yield of crops and excess eater in crops which can lead to fungus and bacteria.
5. Fresh water produce affected badly
6. Biodiversity loss
7. Reduction of human activities
8. Failure in economic development and infrastructure

Weather systems that affect rain pattern include the following:

1. **Air temperature and pressure:** Temperature has a significant impact on rainfall. Increased temperature causes more evaporation, which increases the amount of water vapor in the atmosphere, which, when it achieves a specific amount (assuming all other parameters are stable), induces rainfall. Low pressure causes airmass alignment and lifting, resulting in rainfall.
2. **Humidity:** The proportion of water vapor held by a given volume of air compared to the maximum amount that it can hold. When the humidity is at 80%, the air holds 80% of the total amount of water that it can hold at that temperature. What happens if the humidity exceeds 100%? Excess water condenses and precipitates as ice.
3. **Cloud cover and precipitation:** Clouds affect weather by blocking solar radiation from hitting the ground, capturing heat re-emitted from the ground, and acting as an origin of precipitation. There is far less insulation when there are no clouds. As a consequence, cloudless days can be scorching hot, while cloudless nights can be freezing cold. As a result, cloudy days have a narrower temperature range than clear days.
4. **Wind speed and direction:** All of these are directly proportional to the amount of energy in the system and its location. This energy comes from the sun, which is the ultimate source of it.

The current changing atmosphere correlated with a global mean temperature reform has been seen to have affected and will continue to affect precipitation in different sections of the world. This development is a global challenge that is still considered as gray area. There are several rain models that have been propounded in this regard and the extensive effect such as flood. Changes in precipitation are sometimes tied to the event of desertification and dry spell attributable to precipitation abnormalities. The increasing rainfall in the coastal region and the rise in sea level is due to the melting of the polar ice and the rise in global warming which result in coastal flooding. On the other hand, some parts of the world is experiencing ecological destabilization and disappearance of vegetation caused by reduced rainfall and warmer temperatures which is expected to continue. Below is the rainfall pattern over Africa for over 20 years. This pictorial illustration further supports high and low rainfall tendencies over Africa. From Fig. 3.3, it is interesting to consider Nigeria because it has four different rain patterns.

Lagos (south-west Nigeria) expresses minimal measure of rainfall which happens in the period of December. January is considered as the driest month in Lagos. There is high degree of rainfall between March–October. June is considered as the pinnacle of the rainfall having a value of 386 mm. The month-to-month rainfall in

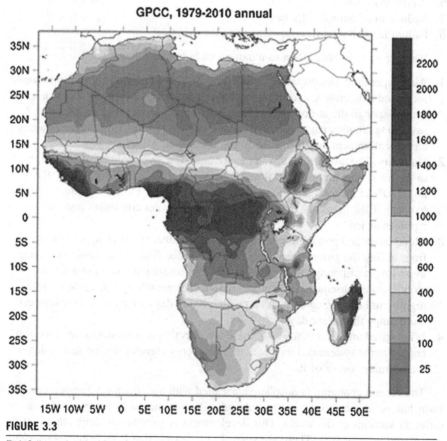

FIGURE 3.3

Rainfall trend over Africa.

Port Harcourt (south-south) increase from March to October and starts diminishing from November to February which is known as dry season. A yearly normal rainfall measure of about 200.45 mm was recorded in Port Harcourt. Rainfall in Port Harcourt shows twofold maxima system with its top in July and September separately. Port Harcourt experience just about 14 days space of practically no rainfall inside July and August. This period is prevalently called the August break. Rainfall in Kano (north-west Nigeria) span from May to mid-October with its top in August while the dry season is from mid-October to mid-May of the following year. Its mean yearly rainfall is between 800 and 900 mm, and the varieties about the mean qualities every year are up to ±30%. Abuja (north–central Nigeria) encounters three diverse climate seasons that have yearly rainfall between 1100–1600 mm.

In engineering, the rainfall–runoff (RR) models are important tools for planning, design and management of water resource systems. The most common models used for RR analysis is the spatial and temporal representations (Chu et al., 2009), while event-based (EB) and continuous simulation (CS) models are the most recognized category within the temporal domain (Elga et al., 2015). The spatial distribution of rainfall is obtained using radiolocation measurement (Jakubiak et al., 2014). Satellite data is also used in this regard because it has spatial character and operative data retrieval.

For any of the rainfall model, rainfall or streamflow event is very vital. Hence, the study of rainfall has wide application in real scenario. In CS modeling, conditions such as initial soil-moisture status, stream and reservoir level, and water table depth (Wu et al., 2013) while EB is dependent on hydrograph (Green and Stephenson, 1986). Despite the kind of analysis of the CS and EB model, the goodness-of-fit is salient for determining model performance. Sharif et al. (2019) highlighted some selected goodness-of-fit is salient for determining model performance as presented in Table 3.1.

1.3 Recent environmental crisis and the problem of climate change

Climate change is the shift in the atmosphere's long-term pattern of activity over time as a result of man-made or natural events. Growing temperatures, wildfires, and drought are becoming more abnormal, weather patterns are shifting, glaciers and snow are melting, and sea levels are rising as a result of climate change. Controlling man-made pollution is essential to mitigate the effects of climate change. Climate change and environmental degradation are some of the biggest problems humanity. Emissions and overexploitation of natural resources are the main causes of this problem, causing global temperatures to rise. Global rise in temperature already present a number of problems which include; rise in sea levels, high thermal discomfort, increased natural disaster cases such as droughts, floods, and bushfires. These effects are expected to worsen over time if action is not taken globally. Looking at the industrialization in the developed world, it is evident that the massive emission of greenhouse gases into the atmosphere. Between 1800 and 2012, the atmospheric CO_2 has increased by about 40% via direct measurements of CO_2 in the atmosphere. The expected changes in climate are based on our understanding of how

Table 3.1 Selected goodness-of-fit measures for model performance (Sharif et al., 2019).

Goodness-of-fit tests	Equation	Interpretation
NSE	$$1 - \frac{\sum_{i=1}^{n}(Q_{obs,i}-Q_{sim,i})^2}{\sum_{i=1}^{n}(Q_{obs,i}-Q_{m,obs})^2}$$	Goodness-of-fit value ranges from $-\infty$ to 1, where 1 means perfect fit and 0 means model validation is as accurate as the observed mean over the observed dataset. Less than 0 means the accuracy of the observed mean over the observed dataset is better than the simulated model.
NSE of daily high flows (ANSE)	$$1 - \frac{\sum_{i=1}^{n}(Q_{obs,i}+Q_{m,obs})(Q_{sim,i}-Q_{obs,i})^2}{\sum_{i=1}^{n}(Q_{obs,i}+Q_{m,obs})(Q_{obs,i}-Q_{m,obs})^2}$$	
NES of daily low flows	$$1 - \frac{\sum_{i}^{n}(\ln(Q_{sim,i}+\varepsilon)-\ln(Q_{obs,i}+\varepsilon))^2}{\sum_{i}^{n}(\ln(Q_{obs,i}+\varepsilon)-\ln(Q_{m,obs}+\varepsilon))^2}$$	
KGE	$1 - \sqrt{(r-1)^2 + (\beta-1)^2 + (\delta-1)^2}$	
R^2	$$\frac{\left[n\sum_{i=1}^{n}(Q_{obs,i}\times Q_{sim,i})-\sum_{i=1}^{n}Q_{obs,i}\times\sum_{i=1}^{n}Q_{sim,i}\right]^2}{\left[n\sum_{i=1}^{n}Q_{obs,i}^2-\left(\sum_{i=1}^{n}Q_{obs,i}\right)^2\right]\times\left[n\sum_{i=1}^{n}Q_{sim,i}^2-\left(\sum_{i=1}^{n}Q_{sim,i}\right)^2\right]}$$	Value ranging from 0 to 1 where 0 means best fit.
RSR	$$\frac{RMSE}{STDEV_{obs}} = \sqrt{\frac{\sum_{i=1}^{n}\left(Q_{sim,i}-Q_{obs,i}\right)^2}{n}\bigg/\sum_{i=1}^{n}(Q_{obs,i}-Q_{m,obs})^2}$$	Value ranges from 0 to $+\infty$, with 0 means perfect fit.
Absolute volume error	$$1 - \frac{\sum_{i=1}^{n}Q_{sim,i}}{\sum_{i=1}^{n}Q_{obs,i}}$$	Value ranging from 0 to $+\infty$, where 0 means best fit and greater than zero means the model underestimates or overestimates overall flow volume or peak flow.
Absolute error in peak flow rate	$1 - \left|\frac{Q_{sim,max}}{Q_{obs,max}}\right|$	

greenhouse gases trap heat and the corresponding increase in global surface temperature i.e., leading to global warming. The rise in CO_2 is largely from combustion of fossil fuels, though significant amount can be ascribed to volcanic eruption, deforestation, biomass burning, and rock weathering. The report from the Royal Society and the US National Academy of Sciences shows that the CO_2 concentrations and temperatures are directly proportional to each other as inferred from less direct methods (see Fig. 3.4). Though people living in developing countries have minima emissions of greenhouse gases into the atmosphere, the realities of ozone depletion

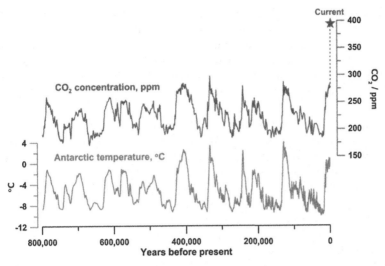

FIGURE 3.4

Evidence that CO_2 concentrations and temperatures are directly proportional (RS and USNAS).

and climate change is universal. Hence, the proposal of models for the healing of the ozone layer is apt for all a sundry. Ozone healing is salient for curbing the effects of climate change. Among them, children, women and people with disabilities are the most exposed while being the least equipped to adapt to a changing climate.

Scientists believe that global temperatures will continue to rise in the coming decades, owing in large part to man-made greenhouse gas emissions. The IPCC (Intergovernmental Panel on Climate Change), which comprises over 1300 scientists from the US and other nations, predicts a temperature increase of 2.5–10°F in the next century. The magnitude of climate change impacts on individual regions may differ with time and with the capacity of various social and environmental processes to mitigate or adapt to change, according to the IPCC.

Recent evidence of climate change include wild fires, drought, shrinking glacier or ice, sea level rising, increased temperature etc. Wildfire risk and severity have risen globally but it is clearly evident in some parts of the world such as Austria and Western United States. Temperature, soil moisture, and the prevalence of trees, shrubs, and other potential fuel are all factors that aids wildfire spread. All of these variables are linked to climate fluctuations and transition, either directly or indirectly. Climate change has increased the drying of organic matter in forests (the fuel that burns and spreads wildfire), doubling the number of major fires in the western United States between 1984 and 2015. Warmer temperatures and drier weather can help fires spread and make them difficult to put out once they start—people are responsible for more than 80% of wildfires in the United States. Warmer, drier conditions often help the spread of the mountain pine beetle and other insects that weaken or destroy trees, accumulating forest fuels.

Climate change, according to research, results in warmer, drier weather. Drought and a longer fire season are leading to the rise in wildfire risk. For most of the western United States, estimates indicate that a 1-degree Celsius rise in average annual temperature will increase the median burned area per year by 600% in some parts of the globe (Ref). In the southeastern United States, modeling indicates increased fire risk and a longer fire season, with the region burned by lightning-ignited wildfire rising by at least 30% from 2011 by 2060.

Drought is affected by a range of variables that are influenced by global climate change. Increased temperatures are likely to result in more precipitation falling as rain instead of snow, fast snow melt, increased evaporation and transpiration etc. Due to earlier snowmelt and warmer temperatures, water supply may become progressively unavailable i.e., leading to high water demands. Since records began in 1895, average drought conditions have varied throughout the world. Droughts were most frequent in the 1930 and 1950s, although the last 50 years have been wetter than normal. Regional patterns, on the other hand, differ. According to a more comprehensive index developed recently, approximately 20—70% of the United States' land area encountered conditions that were at least abnormally dry at any given time between 2000 and 2015.

Shrinking glaciers and Ice is one of the evidence of climate change. The density of the Greenland and Antarctic ice sheets has shrunk. Greenland lost an average of 279 billion tons of ice per year between 1993 and 2019, according to NASA's Gravity Recovery and Climate Experiment, while Antarctica lost around 148 billion tons per year (Ref). In the Alps, Himalayas, Andes, Rockies, Alaska, and Africa, glaciers are retreating almost everywhere. The amount of spring snow cover in the Northern Hemisphere has declined over the last five decades, according to satellite measurements, and the snow is melting sooner. Over the past few decades, both the size and thickness of Arctic sea ice have steadily decreased (Ref). Since 1950, the number of record high temperature events has increased in the United States, while the number of record low temperature events has decreased. In addition, the United States has seen a rise in the number of extreme rainfall events. There are different baselines used to measure global warming temperature. The two most common baselines are (a) pre-industrial conditions, as determined by a 30-year average, and (b) the average from 1980 to 1999. Many public debates about limiting global warming use the former as a starting point. Changes in the amount of water accumulated on land often cause variations in sea level.

Widespread shifts in weather patterns are linked to rising global average temperatures. Extreme weather conditions such as heat waves and massive storms are expected to become more frequent or violent as a result of human-caused climate change, according to scientific reports. Over land areas around the world will experience an increasing annual precipitation to an average rate of 0.08 inches per decade since 1901. Changing weather patterns, on the other hand, have resulted in less precipitation in some regions.

References

Balbis-Morejón, M., Rey-Hernández, J.M., Amaris-Castilla, C., Velasco-Gómez, E., San José-Alonso, J.F., Rey-Martínez, F.J., 2020. Experimental study and analysis of thermal comfort in a University Campus Building in Tropical Climate. Sustainability 12 (21), 8886. https://doi.org/10.3390/su12218886.

Chu, X., ASCE, A.M., Steinman, A., 2009. Event and continuous modelling with HEC-HMS. J. Irrigat. Drain. Eng. 135, 119−124.

Ecophon, 2020. Thermal Comfort. https://www.ecophon.com/in/about-ecophon/functional-demands/thermal-comfort.

Elga, S., Jan, B., Okke, B., 2015. Hydrological modelling of urbanized catchments: a review and future directions. J. Hydrol. 529, 62−81.

Green, I.R.A., Stephenson, D., 1986. Criteria for comparison of single event models. Hydrol. Sci. J. 31, 395−411.

Jakubiak, B., Szturc, J., Ośródka, K., Jurczyk, A., 2014. Experiments with three-dimensional radar reflectivity data assimilation into the COAMPS model. Meteorol. Hydrol. Water Manag. 2 (1), 43−54.

Ličinaa, V.F., Cheung, T., Zhang, H., et al., 2018. Development of the ASHRAE global thermal comfort database II. Build. Environ. 142, 502−512.

Limbachiya, V., Vadodaria, K., Loveday, D., Haines, V., 2012. Identifying a suitable method for studying thermal comfort in people's homes. Loughborough University. Conference contribution. https://hdl.handle.net/2134/11845.

RS, USNAS, 2020. Climate Change Evidence & Causes, an Overview from the Royal Society and the US National Academy of Sciences, pp. 1−22.

Sharif, H., Hewa, G.A., Wella-Hewage, S., 2019. A comparison of continuous and event-based rainfall−runoff (RR) modelling using EPA-SWMM. Water 11, 611.

Wu, J.Y., Thompson, J.R., Kolka, R.K., Franz, K.J., Stewart, T.W., 2013. Using the storm water management model to predict urban headwater stream hydrological response to climate and land cover change. Hydrol. Earth Syst. Sci. 17, 4743−4758.

Generating environmental data: Progress and shortcoming

4

1. Method of generating environmental data: common challenges, safety, and errors

Environmental data is that which is based on the measurement of environmental parameters, and its impacts on the ecosystems. Environmental data is a set of quantitative, qualitative, or geographically referenced facts about the state of the environment and how it is changing.

Quantitative environmental data, statistics, and indicators are commonly spread through databases, spreadsheets, compendia, and other means. Qualitative environmental data consists of descriptions (e.g., textual, pictorial) of the environment or its constituent parts that can't be adequately represented by accurate measurements. Descriptors that are either quantitative or geographically based. Environmental data with a geographic context uses digital maps, satellite imagery, and other sources to provide facts about the environment and its component connected to a map feature or a location. Large volumes of unrefined environmental observations and measurements are based on the environmental data (or its constituents) as well as associated procedures. Environmental dataset can be combined using different types of resources to compile such as the following:

- Surveys of statistical data (censuses or samples) questionnaires.
- Records, registers, and other administrative documents.
- Network monitoring, remote sensing field studies, and published work.

Environmental statistics is a structure organized and grouped according to statistical methods.

Statistical methods are used to analyze environmental data and can be used to infer errors in the measurement procedures and standards. Environmental statistics convert raw data into useful statistics that describe the state and trends of an environmental parameter, as well as the major processes that affect it. For example, Framework for the Development of Environment Statistics (FDES) is a framework for identifying environmental data that falls under its authority. The FDES help with data structuring, organizing, grouping data into statistical series and indicators to generate environmental statistics series. Environment statistics also compile, gather, validate, interpret, and structure environmental data. Environmental indicators are statistics about the environment that need to be processed and interpreted further. Statistics on the environment are often too numerous and comprehensive to meet the needs of policymakers and the general public. Hence, environmental indicators are handy to

- define objectives, assess current and future direction with respect to complex statistics;
- measure parameter that explains, simplify, and pass information;
- evaluate specific programs;
- show progress and track changes over a period of time in a particular condition or situation;
- evaluate the impact of programs and the messages they convey.

For the identification and arranging of indicators, policy frameworks such as the United Nations Sustainable development indicator framework are used.

Environmental data are confronted by various factors as most of the system of measurement are more of open system than closed system. The common challenges for measuring environmental data includes scale, political influence, and uncertainties.

The monitoring program's geographic scale may not correspond to the optimum scale for monitoring, leaving the agency unable or unwilling to conduct effective monitoring. Policymaking based on unnecessary data collection comes with its own set of risks. Different organizations that benefit from and pay for environmental monitoring dictates the politics of environmental decision-making. Political pressing factor could deter the scientific evaluation and processes. As a result of the lingering vulnerability in observing information, it might likewise disallow an organization from utilizing productive checking information to uphold administrative or administrative changes in acquiring data. Indeed, even in the absence of a dominant political player, vulnerability may allow various partners to continue strategic discussions after the observing system has been established.

One vital aspect of environmental research is the detection of errors. We define error as actions or goals that are verifiably and categorically incorrect from a rational or epistemological standpoint (e.g., legitimate paradoxes, numerical missteps, explanations not upheld by the information, mistaken factual methodology, or examining some unacceptable dataset). Statistical analysis errors include errors that do not consistently lend support to the end results. These may occur if the investigations' fundamental assumptions are not met. Environmental research is logical in nature. For example, the best way to know the changes in an observation is to have a control system. When the control system is compromised, it leads to logical errors that may bringing about incoherent results. Sometimes, logical errors occurs in computational investigation of an environmental problem. For example, if we assume an updraft of water vapor from the earth to the atmosphere, there are many factors that determines the rate of updraft. So, if the right assumptions are not made, it may lead to logical errors. Communication errors do not affect information or methods, because they are flaws in scientific reportage. In the most basic case, Communication could be ecstatic, extrapolating beyond what an investigation can tell us. This skill is very important when a numerical analysis of an environmental scenario is considered. The compendium of these errors leads to the generation of "bad idea."

Environmental data is knowledge derived from the environment. Environmental data are generally provided by environmental protection agencies or environmental science agencies. Environmental law enforcement data may also be considered as environmental data. It is important to provide high-quality environmental data in order to carry out a reliable research. It also involves policies that are sensitive and cost-effective, as well as providing the right details for the correct. In order to comply with environmental experiment procedures, certain requirements in accordance with theoretical principles and obligations must be met. For example, data pre-processing, estimation, and validation are vital procedures to ensure that all precautions or reporting standards are met.

1.1 Data quality and errors

There are many versions on what quality of data should look like. Arondel and Depoutot (1998) suggested three features of data quality i.e., comparability of statistics, coherence, and completeness. These three features are very essential because they address the three "Cs" of research i.e., comparability of statistics refers to confirmable; coherence refers to consistency; completeness refers corroborative tendency with facts or trends. On the other hand, Andersson et al. (1997) postulated that data quality can be verified by accuracy, relevance, timeliness, and accessibility. Based on the above several agencies in some countries are guided by policies that ensures that data generation process is transparent for secondary users to re-use for any purpose. For example, The U.S. Office of Management and Budget (1978) enacted guiding principles that ensure the provision of full information of sources, definitions, and methods used in gathering and analyzing dataset to avoid misuse or misrepresentation by secondary data users. In other words, the policy of openness in providing full descriptions of data, methods, assumptions, and sources of error is a practice that is fostered by statistical agencies around the globe. Some scientists are also conversing data users must have the opportunity to review data methods and data limitations (Citro, 1997).

Five sources of error have been identified by expert, i.e., sampling error, nonresponse error, coverage error, measurement error, and processing error. Sampling error is the most popular error in survey. It is the error associated to a sample rather than an entire population of the dataset. Statistical agencies use probability sample to estimate the standard errors of survey estimates. This error is somehow controlled by different statistical error software that are well programmed to detect such errors. This error is not common to survey alone, it is also visible in large environmental field data generation. For example, environment field measurement along certain transverse. For example, the figure below shows how transverse for measuring

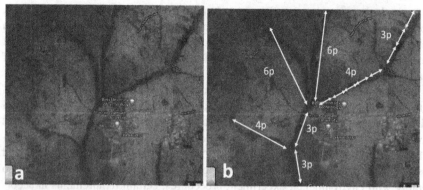

FIGURE 4.1

Field transverses for pollution measurement.

pollution along a river at Ota-Ogun State Nigeria. The transverses are about 17. Should there be a vital error in the measurement as presented in Fig. 4.1, the error is referred to as a sampling error.

Nonresponse error is an error of non-observation reflecting an unsuccessful attempt to obtain the desired information from an eligible unit. Nonresponse reduces sample size, results in increased variance, and introduces a potential for bias in the survey estimates. It is reported using the nonresponse rates. Nonresponse rates may be calculated differently for different purposes (Gonzalez et al. 1994; American Association for Public Opinion Research, 2000) and they are often miscalculated.

Coverage error is the error associated with the failure to include some population units in the frame used for sample selection (undercoverage) and the error associated with the failure to identify units represented on the frame more than once (overcoverage). The source of coverage error is the sampling frame itself. In environmental studies, it ranges from unit conversion error and omission to multiply a range of dataset by a specified number known as the factor. For example, when a liquid environmental sample is characterized using the U-Vis spectrometry. After the intensity is obtained, it is required that it be multiplied by dilution factor. The dilution factor is the ratio of distilled water used to dilute the pure sample collected on the field. Conversion error occurs when there is the need to convert a unit from the measurable unit to the SI unit. A good example of this process can be found in the use of gamma-spec. The conversion is most likely millisievert per year. Another example is the pressure trends from observation stand-point. This type of error are initially minimal but when adopted during numerical procedure could lead to strong trends in pressure.

Measurement error is characterized as the difference between the observed value of a variable and the true, but unobserved, value of that variable. Measurement error comes from four primary sources in survey data collection: the questionnaire, as the official presentation or request for information; the data collection method, as the way in which the request for information is made; the interviewer, as the deliverer of the questions; and the respondent, as the recipient of the request for information. In experiment, the measurement error can arise from the calibration of equipment to the mode of handling equipment. Most equipment have their configuration expiry time which may range from two to about 7 years. After the expiry of the configuration of the equipment, the measurement becomes compromised. The reality of climate change is a good reason for environmentalist not to wait for configuration expiry date. It is advisable to configure the equipment 6 months before the set-time. Another source of measurement error is the handling of the equipment. For example, field officers have reported different height from the ground when using the gamma-spec device. Some suggested 1 meter from the ground, some suggested 40 cm from the ground. These results may significantly vary from each other. Hence, the data is somewhat compromised. The evaluation criteria for the sources of error as declared by US Statistical Policy Office is displayed in Table 4.1.

Processing error occurs after the survey data are collected, during the processes that convert reported data to published estimates and consistent machine-readable information. These errors range from a simple recording error, that is a transcribing or transmission error, to more complex errors arising from a poorly specified edit or imputation model (Morganstein and Marker 1997). The type of error can is visible with secondary user of dataset. If necessary, information about the generation or measurement of dataset is not known. Process errors will be inevitable. Also, the evaluation criteria for background survey information are suggested by the US Statistical Policy Office as presented in Table 4.2. The process error had before now been responsible for the varying results between climate model and obtained data. This error was corrected by the recent correction of the observational database which in turn have presented a perfect correspondence between experimental and computed data (Domingues et al., 2008).

Numerical errors are errors that emerge during the numerical analysis of a theory. These types of errors are basically triggered by minute process error inform of initial values or the type of dataset introduced into the mathematical equations. For example, by neglecting high-order terms in the Taylor expansion, errors are introduced.

In calculus, Taylor's theorem is adjudged as one of the most important basic tools as it is used to obtain the most important single result in numerical computations. It has wide extension and application beyond calculus. It can be incorporated into the Newton or Secant method to examine the convergence of a given environmental dataset. Convergence is an important feature that can be used to know the base line of a process or procedure. Imagine the collection of datasets on pollution emission e.g., PM10 into the environment from a cement factory for a period of 30 days. The average of the dataset is good but may be very misleading when there are cases of spikes from neighboring anthropogenic emission or extreme data low by wind system over the study location. In this case, the baseline which is the average is already compromised. The convergence technique is quite interesting but it requires certain level of skill that will be communicated (in parts) in this book.

Table 4.1 Evaluation criteria for the sources of error (U.S. Statistical Policy Office, 2001).

Error type	Level	Criteria
Coverage error	1	Coverage error is specifically mentioned as a source of nonsampling error
	2	Overall coverage rate is provided Universe is defined Frame is identified and described
	3	Coverage rates for subpopulations are given Poststratification procedures and possible effects are described
Nonresponse error	1	Unit nonresponse is specifically mentioned Item nonresponse is specifically mentioned Overall response rate is given
	2	Item response rates are given Weighted and unweighted unit response rates at each interview level are given Numerator and denominator for unit and item response rate are defined
	3	Subgroup response rates are given Effect of nonresponse adjustment procedure is mentioned Imputation method is described Effect of item nonresponse is mentioned Results of special nonresponse studies are described
Processing error	1	Processing errors are specifically mentioned
	2	Data keying error rates are given Coding error rates are given References are given to processing error studies and documentation
	3	Coder variance studies or other processing error studies are given
Measurement error	1	Measurement error is mentioned as a source of nonsampling error
	2	Specific sources of measurement error are described and defined
	3	Reinterview, record check, or split-sample measurement error studies are mentioned and/or summarized with references to larger reports
Sampling error	1	Sampling error is mentioned as a source of error definition and interpretation of sampling error is included significance level of statements is given
	2	Sampling errors are presented Confidence intervals are defined and method for calculating intervals is described Sampling errors and calculations for different types of estimates (e.g., levels, percent, ratios, means, and medians) are described
	3	Method used for calculating sampling error is mentioned with reference to a more detailed description Generalized model(s) and assumptions are described

Table 4.2 Evaluation criteria for background survey information (U.S. Statistical Policy Office, 2001).

Error type	Level	Criteria
Comparison to other data sources	1	General statement about comparability of the survey data over time is included General statement about comparability with other data sources is included Survey changes that affect comparisons are briefly described
	2	Survey changes that affect comparisons of the survey data over time are described in detail Tables, charts, or figures showing comparisons of the survey data over time are included Tables, charts, or figures showing comparisons with other data sources are included
Sample design	1	General features of sample design (e.g., sample size, number of PSUs and oversampled populations) are briefly described
	2	Sample design methodologies (e.g., PSU stratification variables and methodology, within PSU stratification and sampling methodology and oversampling methodology) are described in detail with references to more detailed documentation
Data collection methods	1	Data collection methods used (e.g., mail, telephone, personal visit) are briefly described
	2	Data collection methods are described in more detail data collection steps taken to reduce nonresponse, undercoverage, or response variance/bias are described
Estimation	1	Estimation methods are described briefly
	2	Methods used for calculating each adjustment factor are described in some detail Variables used to define cells in each step are mentioned cell collapsing criteria used in each step are mentioned

Assume Taylor's theorem where $g(x)$ have $n + 1$ continuous derivatives on $[a,b]$ for some $n > 0$, and let $x, x_o \in [a,b]$.

$$p_n(x) = \sum_{k=0}^{n} \frac{(x - x_o)^k}{k!} f^{(k)}(x_o) \tag{4.1}$$

Eq. (4.1) depicts general functions in terms of polynomials with a known, specified, boundable error. The point x_o is usually chosen at the discretion of the user. In plain numerical studies, x_o is often taken to be 0, however, in environmental studies. x_o may take the value of the minimum value in the measurement, average of the measurement, the median of the given dataset, or converge magnitude of the dataset. When $x_o = 0$, Eq. (4.1) is referred to as Maclaurin series.

In environmental science, Taylor's theorem with remainder is preferred because of the dynamics of data trend during measurement. Then,

$$g(x) = p_n(x) + R_n(x) \qquad (4.2)$$

where $p_n(x)$ and $R_n(x)$ are defined as

$$R_n(x) = \frac{1}{n!} \int_{x_o}^{x} (x - t)^n f^{(n+1)}(t) dt$$

1.2 Satellite measurement

Satellite measurement of environmental parameters e.g., climate, are achieved using the passive and active sensors of both geostationary and polar-orbiting satellites. Satellite missions are characterized by a wide imaging swath and image integration algorithms which makes them fashionable over many decades for data sources in land monitoring and conservation (Kuenzer et al., 2014). The space-based component of the Global Observing System (GOS) for the measurement of environmental and meteorological data includes two constellations: operational geostationary (GSO) and operational non-geostationary (NGSO) low Earth-orbiting, mostly polar-orbiting observation satellites. One of the functions of a GSO MetSat is the collection of environmental data from Data Collection Platforms (DCP). Satellite observation are generated on different scales i.e., individual, regional, and short-term observations to multiple, global, and long-term observations. Multispectral radar, global digital elevation models and radar missions are unique global open data satellite missions of interest for land monitoring and conservation (Dorijan et al., 2020). An extension of the environmental satellite is the climate data record (CDR) (NRC, 2004). CDR is a time series measurement that gives vital information on climate variability and change. Fundamental climate data record (FCDR) is referred to as a time series of inter-calibrated family of sensors to measure geophysical parameter such as sea ice and clouds. Yang et al. (2016) gave a holistic classification of various types of CDR products in a sustainable system as presented in Fig. 4.2. The terms of the parameter in Fig. 4.2 is defined where ICDR is the interim climate data record, FCDR is the fundamental climate data record, TCDR is the thematic climate data record, and CIR is the climate information record. The satellite observations with the aid of retrieved algorithms developed for Environmental Data Record (EDR) of TCDR from FCDR through reprocessing. Till date, there are several challenges that requires the synergy between experts in the climate community and experts in data management only but experts in numerical analysis that could calculate the bias corrections using theories and ground measurement.

A simple way to understand how environmental data are generated is to consider the end-to-end production chain from FCDR through CIR. For example, Heidinger

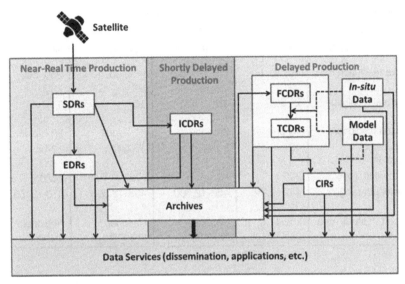

FIGURE 4.2

Schematic diagram of production pathways for various climate data record (CDR) products (Yang et al., 2016).

et al. (2010) reported that the cross- and inter-satellite calibrations can be used to obtain the Advanced Very High Resolution Radiometer (AVHRR) and Pathfinder Atmospheres-Extended (PATMOS-x) reflectance FCDR at the top of the atmosphere (TOA). Heidinger et al. (2012) used same technique to derive the cloud fraction CDR. Zhao et al. (2004) used AVHRR, PATMOS-x reflectance FCDR, and a cloud fraction CDR to generate the aerosol optical thickness (AOT) TCDR.

Multispectral satellite missions make use of the multispectral sensors that works with an algorithm that uses spectral bands based on sensor sensitivity to capture objects on the earth. This remote sensing technique is very much reliable as the image or data retrieval undergoes specific algorithms that ensure image or data detection, identification, and integration are based on differences of spectral signatures per spectral band. A good example of the multispectral satellite missions is the Sentinel (1 & 2), Landsat (7 & 8), MODIS, MERRA etc. (Table 4.3). Sentinel-2 belongs to the ESA Copernicus program. It has moderate spatial resolution with Multispectral Imager (MSI) instrument mounted on two satellites, Sentinel-2A, and Sentinel-2B. It is a polar-orbiting satellite that measures earth's vegetation, land cover, and natural disasters between 84° N and 56° S latitudes (Sentinel, 2021).

Landsat 8 belongs to National Aeronautics and Space Administration (NASA). It has low spatial resolution (based on individual sensors) and better spectral resolution because of the attachment of two multispectral sensors (i.e., Operational Land Imager (OLI) and Thermal Infrared Sensor (TIRS)) that uses specified spectra bands per operation. For example, OLI uses nine spectral bands while TIRS has two

Table 4.3 Properties of selected multispectral satellite missions (Dorijan et al., 2020).

Mission properties	Sentinel-2	Landsat 7	Landsat 8	MODIS
Spatial resolution (m)	10, 20, 60	(15), 30, 60	(15), 30, 100	250, 500, 1000
Temporal resolution (days)	2–3	16	16	1–2
Spectral resolution	13 bands	8 bands	11 bands	25 bands
Radiometric resolution	12-bit	8-bit	16-bit	12-bit
Swath width (km)	290	185	185	2330
Wavelength range (nm)	442–2186	450-12,500	433-12,500	459–2155
Supported study area scale	Local, national	National, regional	National, regional	Regional, global

thermal bands that compliments each other. The sensor upgrade has led to the transition from Landsat 7 to Landsat 8 technology (Landsat, 2021). MODIS also belongs to NASA. It has a multispectral sensor mounted on two satellites i.e., Terra and Aqua. Both satellites orbit diametrically opposite sides of the same polar orbit to generate high temporal resolutions. MODIS observation are intense as it measures several environmental parameters within land, oceans, and lower atmosphere. The upgrades in the MODIS program have led to several products that can be accessed online (MODIS, 2021).

National Oceanic and Atmospheric Administration (NOAA) missions of environmental assessment, prediction, and stewardship is quite significant in environmental data generation. NOAA's four primary mission goals include ecosystems, climate, weather and water, and commerce and transportation. NOAA National Environmental Satellite, Data, and Information Service (NESDIS) is described to be the arm of NOAA that focuses on applications research and development. An important unit of NOAA is the Satellite Oceanography Division (SOD) which has the following functional parts i.e., satellite ocean sensors, ocean dynamics/data assimilation, and marine ecosystems/climate to generate dataset on sea-surface temperature (SST), sea-surface height (SSH), sea-surface roughness (SSR), ocean color, ocean surface winds, and sea ice (Cheney, 2001; Sandwell and Smith, 2001; Clark et al., 2002; Jelenak et al., 2002). In 1987, the CoastWatch Program was established as a unit of SOD. It provides near-real-time environmental satellite data products e.g., sea surface temperature (SST) from the Advanced Very High Resolution Radiometers (AVHRR) on NOAA's polar orbiting spacecraft has comprised the core data stream of NOAA CoastWatch. SOD works in conjunction with NOAA, and other environmental agencies as it does not normally undertake operational activities itself.

Due to recent interest in assimilation, there was the need for the formation of the Joint Center for Satellite Data Assimilation (JCSDA) which is expected to service NOAA, Navy, Air Force, and NASA. Data assimilation has become a principal focus for environmental modeling to aid numerical environmental prediction, including climate investigations.

The Indian Space Research Organization Government of India had shown exceptional breakthrough in space mission. They have 178 missions comprising of 10 student satellites, five experimental missions (one of the missions is dedicated to re-entry of environmental experiment), 31 satellites onboard (for Cartographic applications, urban & rural applications, coastal land use & regulation, utility management like road network monitoring, water distribution, creation of land use maps, Land Information System, GIS applications), 28 customer satellites and 103 satellites for remote sensing, communication, navigation, and space science.

The Japanese space program started in 1955 at the University of Tokyo, where sounding rockets were tested. In 1964 the Japanese space program metaphorized into Institute of Space and Aeronautical Science (ISAS). The period from 1966 to 1969 saw four failed attempts by ISAS to launch Japan's first satellite. The summary overview of the evolution of Japan's space program is shown in Table 4.4.

Numerical Weather Prediction (NWP) is the basis of all modern global and regional weather forecasting. The data generated by the instruments carried by the latest NGSO MetSat systems can be assimilated directly into NWP models to compute forecasts ranging from a few hours up to 10 days ahead. Data obtained by the visible, near-infrared and infrared imagers and other sensors on board GSO MetSat satellites provide input to weather models. The parameter includes sea surface temperature, winds, precipitation estimates, analyses of cloud coverage, height, temperature, solar imagery, etc.

1.3 Modeling procedure

Generally, the steps to developing models include conceptualization, formulation, computational representation, solution, calibration, validation, and application. There are many structures toward developing or building a model which depends on the modeler prerogative. In Eykhoff (1974), two main classification of models were given as priori structural knowledge and the posteriori measurement knowledge. The priori structural knowledge is simply a deductive reasoning approach which is based on existing general theory while the posteriori measurement knowledge is the inductive reasoning approach based on measurement. Both procedures can be combined as presented in Fig. 4.3.

The conceptualization stage is the most important step of modeling because all the groups of parameters, the sequence of the parameter, the conditions at which a parameter takes preeminence, the type of methods that would accommodate all the steps, and the working knowledge of failed models in the past. The latter seem to be the most important point in model conceptualization. There are lots of models in environmental science that are regarded as failed models based on so many factors

Table 4.4 Summary overview of the evolution of Japan's space program.

Timeframe	1955–69	1970s	1980s	1990–2003
Key Achievements	• Establishment of ISAS, NASDA • 1969 agreement allowing transfer of unclassified U.S. launch vehicle technology to Japan	• First successful satellite launch • N-1, N-2 launchers developed • Teaming with U.S. firms to develop satellite capabilities	• Increased Japanese input on launchers (H-1) and satellites • Initiation of H-2 program for indigenous launcher • Agreement to participate in space station • First remote sensin satellite	• Japan achieves independent space capability • Major development programs for launchers and for space station • Military/intel space program initiated
Key Issues	• 1969 agreement prohibited re-exporting launchers	• Dependant on U.S. firms for space capabilities	• Approaching independent status • Forced to open domestic satcom market to competition (lose protected market for Japanese aerospace firms)	• Series of failures in satellite and launcher programs • Japanese firms not successful in commercial competitions • China challenges Japan's position as leading Asian space program • Space program reorganized • Policy review underway (2004)

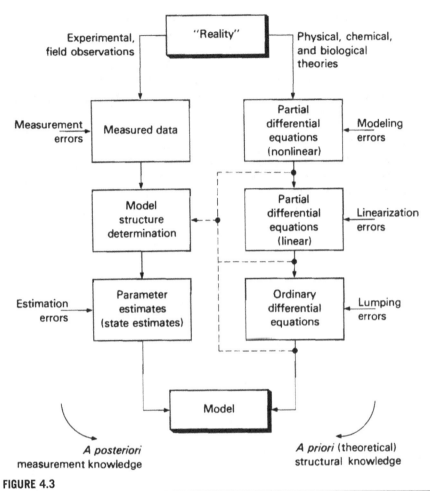

FIGURE 4.3

Combining a priori and a posteriori knowledge in the modeling procedure (Eykhoff, 1974).

peculiar to the problem to be solved. There is no point reinventing the wheel if adequate knowledge of existing models is not acquired. Modeler are expected to understudy existing and failed models to guide the choice of parameters. In some cases, the model based on regional geographical climates would add a parameter that may not be relevant in another geographical region. For example, including snow melting rate in a model would naturally be a failed option for scientist in the tropics. In other words, the sound knowledge of the group parameters would help in christening the model after implementation stage. For example, it would be better to christen a model as "one-dimensional tropical hydrodynamic model" than to give a generic title that may be misleading to researchers from other climatic regions of the globe. The sequence of the parameter, and the conditions at which a parameter takes preeminence is very important in conceptualization this would help the modeler to know exactly what kind of method that would be appropriate to build the model. As discussed earlier, the methods may be in form of a mathematical model (e.g., advection-diffusion water quality model, finite-difference model), computational models (machine learning, simulations, big data analytics), observation models (ecological model, hydrodynamic model) and statistical models. The other elements of a model can be summarized in Fig. 4.4.

Orlob (1982) defined the types of models as "research" or "management" models; internally descriptive (mechanistic) or black box (input-output) models; distributed or lumped models; dynamic or steady state models; stochastic or deterministic models; and nonlinear or linear models. Distributed model has variations in the three orthogonal directions (x, y, z) i.e., 3D system. It is a robust model that have been suggested for environmental problems. It is also the most difficult form of

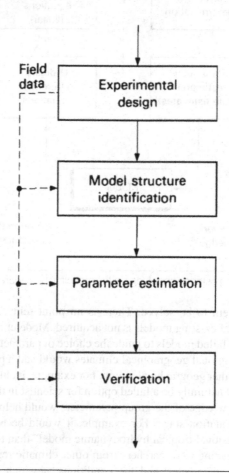

FIGURE 4.4

Calibration and verification phase of the modeling procedure (Orlob, 1982).

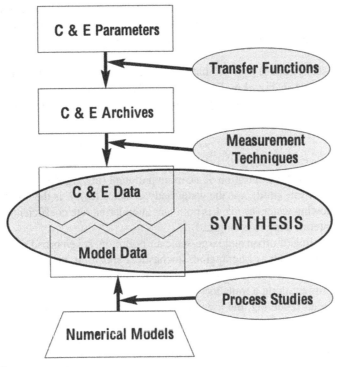

FIGURE 4.5

The role of climate modeling in climate science (Stocker, 2016).

model to solve. The distributed or lumped models are nonlinear models that requires many powerful techniques available for comprehensive analysis of such model. Climate processing procedure propounded by Stocker (2016) is described in Fig. 4.5 below. C & E stands for climate and environmental. The nine processes in the model formulation includes determination of necessary parameters, defining the functions that relates each parameter, archiving different resulting products of the parameters, formulate procedure for gathering data, collect dataset, tweak necessary functions to obtain results from model, compare the model and observation dataset, calculate the biases by synthesizing model and observation dataset.

Fig. 1.3 visualizes the role of modeling in paleoclimate science in a schematic way. Climate change alters certain climate and environmental (C & E) parameters which then can be "read" using appropriate transfer functions. Even in this case, model formulation and application play a central role, but the term climate modeling is not applicable. Climate archives can only be made accessible to research by reliable measurement techniques. An experimental physicist produces climate data (e.g., the reconstruction of the atmospheric CO_2 concentration over the past 800,000 years). The modeler works on the development and application of models that yield model results within the framework process studies. The goal is the synthesis of model results and climate data, which is achieved when the underlying mechanisms and hypotheses are in quantitative agreement. Hence, the model yields a quantitative interpretation of the evolution of climate, based on the laws of physics and chemistry.

Wan and Zeng (2018) gave format of mathematics modeling for water environment as presented in Fig. 4.6. The hydraulic characteristics of water quality was achieved using one-dimensional push-flow model and mass conservation principle as shown below:

$$\frac{\partial c}{\partial t} + \mu \frac{\partial c}{\partial x} = E \frac{\partial^2 c}{\partial x^2} - kc + R$$

where c is the water concentration of a certain pollutant (mg/L), T is the time (d), μ is the average flow rate (m/d), x is the water body distance (m), E is the dispersion coefficient of flowing water (m^2/d), k is pollutant attenuation rate coefficient (d-1), R is the system internal factors is (mg/L.d).

In the modeling of urban and large-scale air pollution, seven broad modeling approaches are used, such as Lagrangian, stochastic, deposition, plume-rise, meteorological, Gaussian, and Eulerian modeling. In Lagrangian modeling, an air parcel (or "puff") is followed along a trajectory, and is assumed to keep its identity during its path. In Eulerian modeling, the area under investigation is divided into grid cells, both in vertical and horizontal directions. Lagrangian modeling are used mostly in Europe over large distances and longer time-periods while Eulerian grid modeling is predominantly used in the US over urban areas.

Meteorological models are developed using two approaches i.e., observation metrological model and numerical meteorological model. The observation metrological model entails understanding local, regional, or global meteorological phenomena. Numerical meteorological models can be divided into two groups i.e., diagnostic

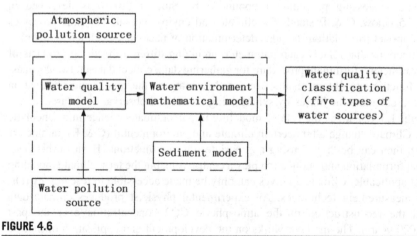

FIGURE 4.6

Format for mathematics modeling of water environment (Wan and Zeng, 2018).

models and prognostic models. Diagnostic models are processes that deals with interpolating and extrapolating available meteorological measurements and contain no time-tendency terms; and prognostic models are models with full time-dependent equations. Plume-rise models deals with both semiempirical and advanced plume-rise formulations. The Gaussian plume model is based on the assumption that the plume concentration, at each downwind distance, has independent Gaussian distributions both in the horizontal and in the vertical. Unlike plume-rise models, Gaussian models deal with the splitting the plume into a series of elements with the inclusion of special dispersion cases. Dispersion models compute the contribution of a source to a receptor as the product of the emission rate multiplied by a dispersion factor. Stochastic models are based on statistical or semiempirical techniques to analyze trends, periodicities, and interrelationships of air quality and atmospheric measurements and to forecast the evolution of pollution episodes.

1.4 **Experimental procedure**

In this section, three main experimental procedures is discussed. Safety is the first rule in any laboratory or field survey. Safety ensures the protection of life and equipment, lowers the cost of maintenance, avoids minor and life-threatening accidents, and prevents major environmental disaster. Common errors in laboratory or field practice were also discussed. This awareness is crafted so that beginner could have adequate knowledge when carrying experiment in the laboratory. Lastly, the maintenance of laboratory apparatus or equipment was discussed.

1.4.1 Safety rules

Basic safety rules for laboratory conduct should be observed whenever working in a laboratory. Same procedure should be observed in field work. Many of the most common safety rules are listed below. Safety rules are divided into safety location rules, pre-experimental safety, experiment safety rules, post-experimental safety, and general safety rules. In this section, the safety rules is discussed in two sections i.e., laboratory and field researchers.

1.4.1.1 Safety for environmental field researcher

The first safety rules in field research are planning and preparation. Most field researcher lacks this skill. Field survey are plagued with many unseen factors ranging from social risk, cultural risk, topographic risk, weather risk, pollution risk, allergy risk, organization risk etc. For example, a researcher who wants to take measurement in a field that have vested interest by big company may likely lose his/her reputation or lives. In this regard, there are uncountable cases of researcher abduction or demise. All this risk must be in the research plan to avoid loss of lives, properties, or reputation.

Based on the above information, it is not advisable for individuals to embark on field research. Group work is highly recommended in this wise. The basis requirement for such group is training on

i. cardiopulmonary resuscitation (CPR) or first aid class;
ii. disease control and prevention;
iii. use of communication devices for local and international rescue program;
iv. use of personal protective equipment;
v. disease vectors catalog.

Office of Environment, Health & Safety (OEHS) University of California, Berkeley, gave a chart that should guide field research across countries as shown in Table 4.5. Another very important aspect of field research is the exposure to certain disease vectors such as bats, rodents, and mosquitoes. In DRC Congo, bats are known as the main carrier of Ebola which had already claimed above a million lives by 2019. The mosquitoes are common in the tropics. Exposure to the anopheles' mosquitoes could lead to malaria. About 700 million people die every year as a result of mosquito-borne disease. The *Aedes aegypti* mosquito transmits zika, dengue, chikungunya, and yellow fever to humans. Exposure to rodents may lead to fatigue, fever, muscle aches, and sometimes headaches, dizziness, chills, abdominal problems, coughing, and shortness of breath.

Contact Cal Adventures (642-

1.4.1.2 Safety for environmental laboratory researcher
1.4.1.2.1 Safety location rules

1) Know locations of laboratory safety showers, eye wash stations, and fire extinguishers.
2) Know emergency exit routes.
3) Laboratory safety glasses or goggles should be worn in any area where chemicals are used or stored. They should also be worn any time there is a chance of splashes or particulates to enter the eye. Closed toe shoes will be worn at all times in the laboratory. Perforated shoes or sandals are not appropriate.
4) Keep all sink traps (including cup sink traps and floor drains) filled with water by running water down the drain at least monthly.
5) Avoid working alone in a building. Do not work alone in a laboratory if the procedures being conducted are hazardous.
6) Familiarize yourself waste storage locations around the laboratory.
7) When the fire alarm sounds you must evacuate the building via the nearest exit.
8) Ensure you are fully aware of your facility's/building's evacuation procedures.
9) Lab areas containing carcinogens, radioisotopes, biohazards, and lasers should be properly marked with the appropriate warning signs.

1.4.2 Pre-experimental safety

1) Never consume and/or store food or beverages or apply cosmetics in areas where hazardous chemicals are used or stored.
2) Long hair and loose clothing must be pulled back and secured from entanglement or potential capture.

Table 4.5 Physical and environmental hazards found worldwide (OEHS, 2021).

Hazard	Location	Cause	Symptoms	First aid	Prevention
Dehydration	Worldwide	Not enough water intake	Dark urine Lethargy Constipation Light-headedness	Drink plenty of fluids, take frequent rest breaks, and minimize intake of beverages containing caffeine.	Drink plenty of water (at least 2 quarts of water per day). Drink more if working strenuously or in a warm climate.
Impure water	Worldwide	Harmful organisms and pathogens living in "natural" water sources	Gastrointestinal illness Flu-like symptoms	Drink clear liquids. Slowly introduce mild foods, such as rice, toast, crackers, bananas, or applesauce. See a doctor if there is no improvement.	Carry your own water. Treat water before use with tablets, purifiers, or by boiling for more than 3 min.
Sunburn	Worldwide	Excessive exposure to the sun	Irritated skin, pink or red in color	Apply cool water, aloe, or other cooling lotion to affected area.	Wear long sleeved clothing and a hat. Apply sunblock with sun protection factor (SPF) of 30.
Heat Exhaustion	Worldwide: hot climates	Prolonged physical exertion in a hot environment	Fatigue Excessive thirst Heavy sweating Cool and clammy skin	Cool the victim, treat for shock, and slowly give water or electrolyte replacer.	Acclimate to heat gradually. Drink plenty of liquids. Take frequent rest breaks.
Heat Stroke	Worldwide: hot climates	Prolonged physical exertion in a hot environment	Exhaustion Light-headedness Bright red skin which is warm to the touch	Cool the victim at once, replenish fluids, and seek medical attention immediately.	Acclimate to heat gradually. Drink plenty of liquids. Take frequent rest breaks.

Continued

Table 4.5 Physical and environmental hazards found worldwide (OEHS, 2021).—cont'd

Hazard	Location	Cause	Symptoms	First aid	Prevention
Frostbite	Worldwide: cold climates	Exposure to cold temperatures	Waxy, whitish numb skin Swelling, itching, burning, and deep pain as the skin warms	Slowly warm the affected areas (do NOT rub area) and seek medical attention as soon as possible.	Dress in layers. Cover your extremities with warm hats, face mask, gloves, socks, and shoes.
Hypothermia	Worldwide: cold climates	Prolonged exposure to cold temperatures	Shivering Numbness Slurred speech Excessive fatigue	Remove cold, wet clothes. Put on dry clothes or use a blanket or skin-to-skin contact to warm up. Drink warm liquids and seek medical attention as soon as possible.	Dress in layers. Wear appropriate clothing. Avoid getting damp from perspiration.
Carbon Monoxide	Worldwide	Running a vehicle or burning a fuel stove in an enclosed space	Severe headaches Disorientation Agitation Lethargy Stupor Coma	Remove the victim to fresh air immediately and perform CPR if needed.	Keep areas adequately ventilated when burning fuel. Ensure that vehicle tailpipe is not covered by snow.
Extreme weather	Worldwide	Snow squalls, blizzards, lightning, tornadoes, hurricanes, monsoon rains, floods	Severe weather can result in physical injury and/or death	Seek shelter immediately.	Be aware of special weather concerns. Bring appropriate equipment to deal with severe weather.
High Altitude Illness	Worldwide: High altitudes	Decreased oxygen and increased breathing rate	Headache Nausea Weakness	Use supplemental oxygen and decrease altitude.	Allow your body to acclimatize by gaining elevation slowly.

3) Determine the potential hazards and appropriate safety precautions before beginning any work.

4) All equipment should be regularly inspected for wear or deterioration.

5) Clothing made of synthetic fibers should not be worn while working with flammable liquids or when a fire hazard is present as these materials tend to melt and stick to exposed skin.

6) Avoid wearing jewelry in the lab as this can pose multiple safety hazards.

7) Doing experiments in the laboratory without supervision is prohibited. The performance of unauthorized experiments and the use of any equipment in an unauthorized or unsafe manner are strictly forbidden.

8) Coats, bags, and other personal items should be stored in the proper areas; not on the bench tops or in the aisle ways. Beginner and instructors should be able to easily move around lab tables.

1.4.3 Experimental safety rules

1) Horseplay is not tolerated during experiment.

2) Avoid distracting or startling persons working in the laboratory.

3) Minimize all chemical exposures.

4) Do not pour chemicals down drains. Do NOT utilize the sewer for chemical waste disposal.

5) No contact lenses should be worn around hazardous chemicals—even when wearing safety glasses.

6) Perform work with hazardous chemicals in a properly working fume hood to reduce potential exposures.

7) To avoid exposure to injuries, open-backed shirts, bare midriff shirts, or shirts which expose areas of the torso are not permitted.

8) Please exercise caution when dealing with electrical devices.

9) When working with low power lasers (Class 2 lasers, power <1 mW in visible range) never look directly into the beam, and never direct it at another person. Keep the beam at lower level than your eyes, and remove jewelry, watches and other shiny objects during alignment, as they can reflect and accidently direct the beam into another person's eyes. Always wear laser goggles for lasers operating outside the visible wavelength range.

10) If an instrument or piece of equipment fails during use, or isn't operating properly, report the issue to a technician right away. Never try to repair an equipment problem on your own.

11) Make sure you always follow the proper procedures for disposing lab waste.

12) When working with electrical circuits, be sure that the current is turned off before making adjustments in the circuit.

13) Do not connect the terminals of a battery or power supply to each other with a wire. Such a wire will become dangerously hot.

1.4.4 Post- experimental safety

1) Never leave containers of chemicals open.

2) Wash exposed areas of the skin prior to leaving the laboratory.

3) Do not remove apparatus from cabinets without the permission of the instructor.

4) Return all of your equipment and glassware to its original location at the end of the laboratory session.

5) Extinguish all flames and turn off all equipment, as appropriate, before leaving.

6) All personal accidents, injuries and illnesses, however slight, occurring in the laboratory must be reported immediately to the instructor.

7) Any liquid (safe liquid) spills must be cleaned up immediately to avoid injuries. In case of bigger leaks, the appropriate authorities (the primary faculty in charge or other department faculty/staff) must be notified.

8) When using hot plates, unplug them before you leave the lab.

9) Return all equipment, clean and in good condition, to the designated location at the end of the lab period. Leave your lab area cleaner than you found it.

1.4.5 General safety rules

1) Avoid skin and eye contact with all chemicals.

2) Use equipment only for its designated purpose.

3) All containers must have appropriate labels. Unlabeled chemicals should never be used.

4) Do not taste or intentionally sniff chemicals.

5) Do not utilize fume hoods for evaporations and disposal of volatile solvents.

6) Equipment should be maintained according to the manufacturer's requirements and records of certification, maintenance, or repairs should be maintained for the life of the equipment.

7) No cell phone or ear phone usage should be allowed in the active portion of the laboratories, or during experimental operations.

8) Confine loose clothing and tie up long hair while performing experiments.

9) The use of headphones (i.e., iPods) other than approved hearing protection devices is prohibited.

10) Do not eat or drink in the laboratory. Do not use laboratory containers and utensils for food and drinks storage.

11) Visitors, including children, are not permitted to enter the laboratories.

12) Do not take laboratory equipment (including glassware) outside the lab without the permission of the instructor.

13) Always work in properly-ventilated areas.

14) Never use lab equipment that you are not approved or trained by your supervisor to operate.

2. Common errors in laboratory practice

There are the following areas where the mistake can lead to erroneous laboratory results. Hence, there is the need to understand why we must avoid these errors.

1) Random error is said to take place when repeated measurements of the quantity, give different values under the same conditions. It occurs due to some

unknown reasons. Taking several readings of the same quantity and then taking their mean value can reduce the random error.

2) Systematic errors occur when all the measurements of physical quantities are affected equally, these give the consistent difference in the readings. The systematic errors may occur due to (a) zero error in measuring instrument, (b) poor calibration of the instrument, (c) incorrect calibration on the measuring instruments.

3) Avoid zero error in all measuring apparatus. Always reconfigure your apparatus before taking measurement.

4) Observational error that include parallax in reading a meter scale.

5) Unpredictable fluctuations in line voltage, temperature, or mechanical vibrations of equipment.

6) A final source of error, called a blunder, is an outright mistake. A person may record a wrong value, misread a scale, forget a digit when reading a scale or recording a measurement, or make a similar blunder. These blunders should stick out like sore thumbs if we make multiple measurements or if one person checks the work of another. Blunders should not be included in the analysis of data.

7) Scale error is a known laboratory error. If a piece of equipment is not calibrated correctly (e.g., a wooden ruler has shrunk), all measurements will be offset by the same fraction.

8) The most common error in physics is the parallax error. If you make a measurement by comparing an indicator against a scale (e.g., reading a dial on a voltmeter, or using a mercury thermometer), the angle at which you view it will affect the reading.

9) Errors arising from the environment i.e., ideally, the control variables are kept constant, but some may be beyond your control, e.g., air pressure, temperature, humidity, vibrations.

10) Measurement errors from insufficient precision i.e., if you're measuring something that falls between two markings on a scale (e.g., you're using a ruler to measure something that's 10.25 mm long), you cannot measure its precise value and will need to round it up or down (does it look like 10 mm or 10.5 mm?).

11) Calculating Experimental Error is common in manual measuring procedure. When a scientist reports the results of an experiment, the report must describe the accuracy and precision of the experimental measurements. Some common ways to describe accuracy and precision are described below.

3. **Maintaining laboratory apparatus**

The care and maintenance of laboratory equipment is an integral part of quality assurance in the lab. Well-maintained lab equipment ensures that data is consistent and reliable, which in turn impacts the productivity and integrity of the work produced.

In addition, routine maintenance ensures that lab equipment is safe for use through highlighting and repair of faulty equipment and equipment parts.

Various procedures and routines will ensure that your laboratory equipment is well-maintained and cared for; this includes:

1) Developing standard operating procedures for all lab equipment.
2) Preparing documentation on each specific equipment, outlining the repairs and maintenance undertaken.
3) Outlining a preventive maintenance program for each equipment.
4) Clean the equipment using a piece of clean cloth. If there is a stain, a mild detergent can be applied. Also a paintbrush with soft bristles can be used to remove particles or dust deposited on the weight plate.
5) Verify that the adjustment mechanisms apparatus adequately.
6) One point calibration is recommended for some equipment. This is carried out for normal working conditions and for normal use.
7) Two-point calibration is recommended for some equipment. This is done prior to performing very precise measurements. It is also done if the instrument is used sporadically and its maintenance is not carried out frequently.

References

Andersson, C., Lindstrom, H., Lyberg, L., 1997. Quality declaration at statistics Sweden. In: Seminar on Statistical Methodology in the Public Service. U.S. Office of Management and Budget, Washington, DC, pp. 131–144 (Statistical Policy Working Paper 26).

Arondel, P., Depoutot, R., May 1998. Overview of Quality Issues when Dealing with Soci-Economic Products in an International Environment (Paper prepared for presentation at the XXXth ASU Meeting).

Cheney, R.E., 2001. Satellite altimetry. In: Encyclopedia of Ocean Sciences. Academic Press, London, pp. 2504–2510.

Citro, C., 1997. Discussion. In: Seminar on Statistical Methodology in the Public Service. U.S. Office of Management and Budget, Washington, DC, pp. 43–51 (Statistical Policy Working Paper 26).

Clark, D.K., Yarbrough, M.A., Feinholz, M., Flora, S., Broenkow, W., Kim, Y.S., Johnson, B.C., Brown, S.W., Yuen, M., Mueller, J.L., 2002. MOBY, A radiometric buoy for performance monitoring and vicarious calibration of satellite ocean color sensors: measurement and data analysis protocols. In: Ocean Optics Protocols for Satellite Ocean Color Sensor Validation, Rev 3, vol. 2. NASA/TM-2002-210004/Rev3-Vol2.

Domingues, C.M., Church, J.A., White, N.J., Gleckler, P.J., Wijffels, S.E., Barker, P.M., Dunn, J.R., 2008. Improved estimates of upper-ocean warming and multi-decadal sea-level rise. Nature 453, 1090–1093.

Dorijan, R., Obhodaš, J., Mladen, J., Mateo, G., 2020. Global open data remote sensing satellite missions for land monitoring and conservation: A review. Land 9, 402.

Eykhoff, P., 1974. System Identification- Parameter and State Estimation. Wiley, Chichester.

Gonzalez, M.E., Kasprzyk, D., Scheuren, F., 1994. Nonresponse in federal surveys: an exploratory study. In: Amstat News 208. American Statistical Association, Alexandria, VA.

Heidinger, A.K., Straka, W.C., Molling, C.C., Sullivan, J.T., Wu, X.Q., 2010. Deriving an inter-sensor consistent calibration for the AVHRR solar reflectance data record. Int. J. Rem. Sens. 31, 6493–6517.

Heidinger, A.K., Evan, A.T., Foster, M.J., 2012. A Naive Bayesian cloud-detection scheme derived from CALIPSO and applied within PATMOS-x. J. Appl. Meteorol. Climatol. 51, 1129–1144.

Jelenak, Z., Connor, L.N., Chang, P.S., 2002. The accuracy of high resolution winds from QuikSCAT. In: IEEE International Geoscience and Remote Sensing Symposium, Toronto, Canada.

Kuenzer, C., Van Beijima, S., Gessner, U., Dech, S., 2014. Land surface dynamics and environmental challenges of the Niger Delta, Africa: remote sensing-based analyses spanning three decades (1986–2013). Appl. Geogr. 53, 354–368.

Landsat 8 Data Users Handbook, 2021. Available online: https://www.usgs.gov/media/files/landsat-8-data-users- handbook. (Accessed 26 May 2021).

MODIS Surface Reflectance User's Guide, 2021. Available online: https://modis-land.gsfc.nasa.gov/pdf/MOD09_ UserGuide_v1.4.pdf. (Accessed 26 May 2021).

Morganstein, D., Marker, D., 1997. Continuous quality improvement in statistical agencies. In: Lyberg, L., Biemer, P., Collins, M., deLeeuw, E., Dippo, C., Schwarz, N., Trewin, D. (Eds.), Survey Measurement and Process Quality. John Wiley & Sons, New York, pp. 475–500.

NRC (National Research Council), 2004. Climate Data Records from Environmental Satellites: Interim Report. National Academies Press, Washington, DC, USA.

OEHS, 2021. Safety Guidelines for Field Researchers.

Orlob Gerald, T., 1982. Mathematical Modeling of Water Quality: Streams, Lakes, and Reservoirs. A Wiley-Interscience Publication International Institute for Applied Systems Analysis, pp. 1–186.

Sandwell, D.T., Smith, W.H.F., 2001. Bathymetric estimation. In: Fu, L.-L., Cazenave, A. (Eds.), Satellite Altimetry and Earth Sciences. Academic Press, San Diego, CA, pp. 441–457.

Sentinel, 2021. 2 User Handbook. Available online: https://earth.esa.int/documents/247904/685211/Sentinel-2_ User_Handbook. (Accessed 26 May 2021).

Stocker, T., 2016. Introduction to Climate Modeling. Physikalisches Institut Universität Bern, pp. 15–76.

U.S. Office of Management and Budget, 1978. Statistical Policy Handbook (Washington, DC).

US Statistical Policy Office, Office of Information and Regulatory Affairs Office of Management and Budget, 2001. Measuring and Reporting Sources of Error in Surveys, pp. 1–65.

Wan, D., Zeng, H., 2018. Water environment mathematical model mathematical algorithm. IOP Conf. Ser. Earth Environ. Sci. 170, 032133.

Wenze, Y., Viju, O.J., Xuepeng, Z., Hui, L., Kenneth, R.K., 2016. Satellite climate data records: development, applications, and societal benefits. Remote Sens 8, 331.

Yang, W., John, V.O., Zhao, X., Lu, H., Knapp, K.R., 2016. Satellite climate data records: Development, applications, and societal benefits. Remote Sensing 8 (4), 331.

Zhao, X.-P., Dubovik, O., Smirnov, A., Holben, B.N., Sapper, J., Pietras, C., Voss, K.J., Frouin, R., 2004. Regional evaluation of an advanced very high resolution radiometer (AVHRR) two-channel aerosol retrieval algorithm. J. Geophys. Res. 109, D02204.

Root finding technique in environmental research

1. Application of root finding technique to environmental data

The root finding method is a popular method that has little application to science, technology, or engineering. However, it has salient application to all fields of life as far there are dataset involved. In this chapter, the application of this technique in application to datasets and how it can be applied to environmental fields.

1.1 The root finding method

Solve $f(x) - 0$ for x, when an explicit analytical solution is impossible. Sometimes, the solution of the problems sometime is "exact" in a fixed amount of time. The solution of other problems may have error tolerances, and our algorithms may have to iterate to compute them. Generally, the shortfall for the root finding system is its illogical termination of problems despite the sophistication of the algorithms and the duration at which convergence would be achieved at lower error tolerances. However, the advantages far outweigh the aforementioned disadvantages. In this chapter, three main root finding method will be discussed, i.e., Bisection, Secant, and Newton's method. In practice, the duration of converge of these methods is in descending order, i.e., Newton's method, Secant, and Bisection. By principle, the three methods, i.e., Bisection, Secant, and Newton's method are defined in Eqs. (5.1)−(5.3), respectively. Generally, a sequence $x_0, x_1, x_2, \ldots$ are constructed such that it converges to the root $x = r$.

$$x_{n+1} = x_{n-1} + \frac{x_n - x_{n-1}}{2} \tag{5.1}$$

$$x_{n+1} = x_n - \frac{(x_n - x_{n-1})f(x_n)}{f(x_n) - f(x_{n-1})} \tag{5.2}$$

$$x_{n+1} = x_n - \frac{f(x_n)}{f'_{(x_n)}} \tag{5.3}$$

Numerical Methods in Environmental Data Analysis. https://doi.org/10.1016/B978-0-12-818971-9.00009-0

Example 1: Find the root of the function $f(x) = x^2 - 3$ using the Bisection, Secant, and Newton's method.

1.1.1 Newton's method

Let's assume that $x_0 = 1.2$ in Eq. (5.3).

$$x_{n+1} = x_n - \frac{f(x_n)}{f'(x_n)}$$

where $f'(x) = 2x$

$x_1 = 1.2 - \frac{(1.2)^2 - 3}{2*1.2}$
$x_1 = 1.2 + 0.65$
$x_1 = 1.85$

$x_2 = 1.85 - \frac{(1.85)^2 - 3}{2*1.85}$
$x_2 = 1.85 - 0.11$
$x_2 = 1.74$

$x_3 = 1.74 - \frac{(1.74)^2 - 3}{2*1.74}$
$x_3 = 1.74 - 0.01$
$x_3 = 1.73$

1.1.2 Secant method

Let's assume that $x_0 = 1.2$ and $x_{-1} = 0.9$ in Eq. (5.2).

$$x_{n+1} = x_n - \frac{(x_n - x_{n-1})f(x_n)}{f(x_n) - f(x_{n-1})}$$

$$x_1 = 1.2 - \frac{(1.2 - 0.9)((1.2)^2 - 3)}{((1.2)^2 - 3) - ((0.9)^2 - 3)}$$

$$x_1 = 1.34$$

$$x_2 = 1.34 - \frac{(1.34 - 1.2)((1.34)^2 - 3)}{((1.34)^2 - 3) - ((1.2)^2 - 3)}$$

$$x_1 = 1.43$$

$$x_3 = 1.43 - \frac{(1.43 - 1.34)((1.43)^2 - 3)}{((1.43)^2 - 3) - ((1.34)^2 - 3)}$$

$$x_1 = 1.51$$

$$x_4 = 1.51 - \frac{(1.51 - 1.43)((1.51)^2 - 3)}{((1.51)^2 - 3) - ((1.43)^2 - 3)}$$

$$x_1 = 1.59$$

$$x_4 = 1.59 - \frac{(1.59 - 1.51)((1.59)^2 - 3)}{((1.59)^2 - 3) - ((1.51)^2 - 3)}$$

$$x_1 = 1.7$$

$$x_5 = 1.7 - \frac{(1.7 - 1.59)\left((1.7)^2 - 3\right)}{\left((1.7)^2 - 3\right) - \left((1.59)^2 - 3\right)}$$

$$x_1 = 1.7$$

1.1.3 Bisection method

Let's assume that $x_0 = 1.2$ and $x_{-1} = 0.9$ in Eq. (5.1).

$$x_{n+1} = x_{n-1} + \frac{x_n - x_{n-1}}{2}$$

$$x_1 = 0.9 + \frac{1.2 - 0.9}{2}$$

$$x_1 = 1.05$$

$$x_2 = 1.2 + \frac{1.05 - 1.2}{2}$$

$$x_2 = 1.125$$

$$x_3 = 1.05 + \frac{1.125 - 1.05}{2}$$

$$x_3 = 1.088$$

$$x_4 = 1.125 + \frac{1.088 - 1.125}{2}$$

$$x_4 = 1.11$$

$$x_5 = 1.088 + \frac{1.11 - 1.088}{2}$$

$$x_5 \neq 1.099$$

Hence, the two assumptions in the bisection method must be higher than the secant method. In the second trial, the following assumptions were made, i.e., $x_0 = 1.8$ and $x_{-1} = 1.5$ in Eq. (5.1).

$$x_{n+1} = x_{n-1} + \frac{x_n - x_{n-1}}{2}$$

$$x_1 = 1.5 + \frac{1.8 - 1.5}{2}$$

$$x_1 = 1.65$$

$$x_2 = 1.8 + \frac{1.65 - 1.8}{2}$$

$$x_2 = 1.725$$

$$x_3 = 1.65 + \frac{1.725 - 1.65}{2}$$

$$x_3 = 1.69$$

$$x_4 = 1.725 + \frac{1.69 - 1.725}{2}$$

$$x_4 = 1.71$$

From the above exercises, it is clearly affirmed that most numerical solution to an identified problem results in certain errors in the solutions. Hence, before designing the numerical procedures, it is expected that some salient steps to minimize errors are in place. Hence, it is advisable that before running a code of algorithms (as will be seen in the next section), several codes to normalize the parameters are run. In the previous chapters, different computational errors were discussed, but in this section, and in the succeeding chapter, it is important to note that there are two main types of errors associated to numerical solutions, i.e., truncation and roundoff errors. The truncation errors occur as a result of the approximate nature of numerical solutions, i.e., they often consist of lower order terms and higher order terms. For example, in the last solution or the bisection method, x_1, x_2, x_3, and x_4 fluctuated in an oscillating manner before it got to the desired solution. Sometimes, depending on your codes, the latter terms are often ignored in the computations for the sake of computational efficiency, resulting in error in the solution. The second type of error, i.e., round-off error occurs in the computer system. For example, the allowable decimal place (dp) of the computer which most times is 7dp portends any number after the 8th decimal point will be dropped. Hence, if the algorithm requires huge number of operations, a significant portion of the results will be ignored, i.e., leading to compromised process.

1.2 Modification of the root finding method to data application

Nonlinear systems appear in flooding, weather forecasting, heat waves forecasting, population dynamics (demography), water management, disease/infection modeling, chemical reactions, biological reactions, air pollution, pandemic, traffic management etc. Most of the times, a nonlinear system may have mathematical equations in algebraic (polynomials) form or transcendental form. The polynomial equation is very common as it is very easy to generate (i.e., as would be demonstrated later on).

A nonlinear model/equation does not have the general linear form as:

$$y = t_1 x_1 + t_2 x_2 + t_3 x_3 \ldots + t_n x_n \tag{5.4}$$

Hence, it does not satisfy the following conditions:

1.2.1 Additivity (superposition)

$$y(t + x) = y(t) + y(x) \tag{5.5}$$

1.2.2 Homogeneity (linearity)

$$y(t * x) = t * y(x) \tag{5.6}$$

In this section, Eqs. (5.1)–(5.3) will be explained using one of the simplest tools, i.e., Microsoft Excel Package. Then we shall go into a more complex demonstration using MATLAB. In this case, we will be showing how the root finding method cam be used for larger/big dataset.

1.2.2.1 Example 1: performing root finding technique using Microsoft Excel Package

Here, we would demonstrate using live/field dataset as presented in Table 5.1. The dataset below is the measurement of radioactive dose from a field study within 16 locations but only six locations is discussed. The parameters are thorium, potassium, and radon. The demonstration of the determination of the algebraic representation will be demonstrated in pictorial form. The following steps were used to obtain the algebraic representation.

Step 1: Plot the graph for each radioactive dose, i.e., using scattered plot
Step 2: Use trendline to determine the polynomial equation

Table 5.1 Field measurement of radioactive dose.

Location	Th-232 (Bqkg-1)	K-40 (Bqkg-1)	Ra-226
1	13.45	17.29	16.28
2	24.86	8.41	22.37
3	13.74	13.51	12.44
4	21.61	19.53	19.04
5	21.84	37.33	17.16
6	44.29	23.58	30.67

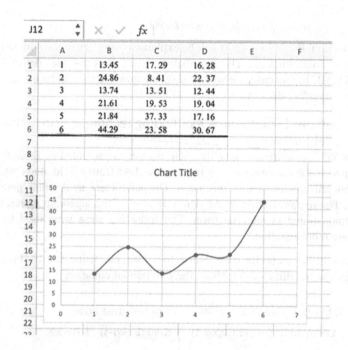

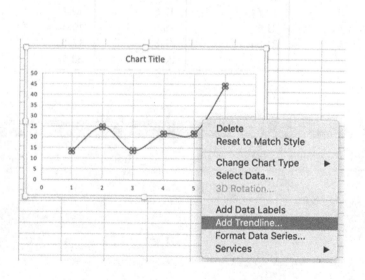

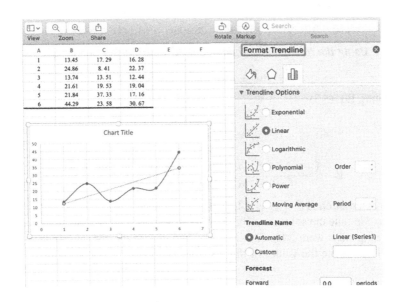

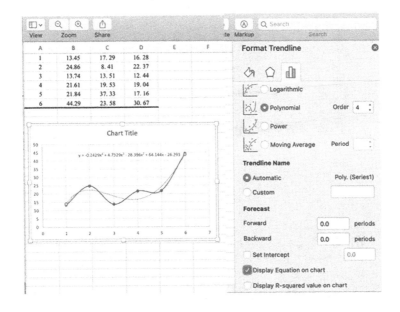

The polynomial equation for the thorium measurement is given as:

$$y = -0.2429x^4 + 4.7329x^3 - 28.396x^2 + 64.144x - 26.293 \qquad (5.7)$$

1.2.2.1.1 Using the Newton's method

$$y' = -0.9716x^3 + 14.1987x^2 - 56.792x + 64.144 \qquad (5.7)$$

Applying Eq. (5.3)

$$x_{n+1} = x_n - \frac{f(x_n)}{f'_{(x_n)}}$$

$$x_{n+1} = x_n - \frac{\left(-0.2429x^4 + 4.7329x^3 - 28.396x^2 + 64.144x - 26.293\right)}{\left(-0.9716x^3 + 14.1987x^2 - 56.792x + 64.144\right)}$$

If the assumption is that $x_0 = 14.8$, then the same Microsoft Excel Package can be used to compute the solutions. The steps to do that are a little bit technical; hence, the following steps were adopted.

Step 1: Compute the values of the first row to obtain the first iteration:

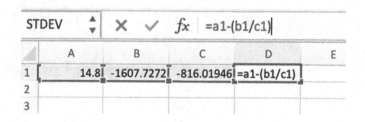

Step 2: Compute the values of the second row to obtain the remaining iterations:

		f_x	=-0.2429*D1^4+4.7329*D1^3 - 28.396*D1^2 + 64.144*D1 - 26.293					
A	B	C	D	E	F	G	H	I
14.8	-1607.7272	-816.01946	12.8297931					
	-463.55775	-379.18264	11.6072747					

		f_x	=-0.9716*D1^3+14.1987*D1^2 - 56.792*D1 + 64.144				
A	B	C	D	E	F	G	H
14.8	-1607.7272	-816.01946	12.8297931				
	-463.55775	-379.18264	11.6072747				

		f_x	=D1-(B2/C2)		
A	B	C	D	E	F
14.8	-1607.7272	-816.01946	12.8297931		
	-463.55775	-379.18264	11.6072747		

		f_x	=-0.2429*D1^4+4.7329*D1^3 - 28.396*D1^2 + 64.144*D1 - 26.293					
A	B	C	D	E	F	G	H	I
14.8	-1607.7272	-816.01946	12.8297931					
	-463.55775	-379.18264	11.6072747					

| 2 | ▲▼ | ✕ | ✓ | *fx* | =-0.2429*D1^4+4.7329*D1^3 - 28.396*D1^2 + 64.144*D1 - 26.293 |

A	B	C	D	E	F	G	H	I
14.8	-1607.7272	-816.01946	12.8297931					
	-463.55775	-379.18264	11.6072747					
	-115.1294	-201.50376	11.0359236					
	-18.372663	-139.23659	10.9039707					
	-0.8429491	-126.56324	10.8973104					
	-0.0020767	-125.9399	10.897294					
	-1.271E-08	-125.93836	10.897294					
	8.0291E-13	-125.93836	10.897294					
	-1.016E-12	-125.93836	10.897294					

Step 3: Confirm that the computation command of the last cell:

| ▲▼ | ✕ | ✓ | *fx* | =D8-(B9/C9) |

A	B	C	D	E
14.8	-1607.7272	-816.01946	12.8297931	
	-463.55775	-379.18264	11.6072747	
	-115.1294	-201.50376	11.0359236	
	-18.372663	-139.23659	10.9039707	
	-0.8429491	-126.56324	10.8973104	
	-0.0020767	-125.9399	10.897294	
	-1.271E-08	-125.93836	10.897294	
	8.0291E-13	-125.93836	10.897294	
	-1.016E-12	-125.93836	10.897294	

If this is done, the root of the graph is 10.897294 and the convergence graph is shown below. The convergence showed that at the root of the dataset was obtained at the fourth iteration.

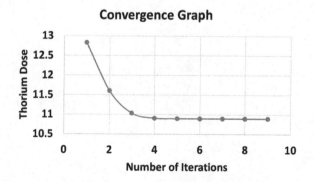

Convergence Graph

1.2.2.1.2 Using the secant method

Using Eq. (5.2)

$$x_{n+1} = x_n - \frac{(x_n - x_{n-1})f(x_n)}{f(x_n) - f(x_{n-1})}$$

If the assumptions are $x_0 = 15.32$ and $x_{-1} = 14.8$, then the Microsoft Excel Package can be used to compute the solutions. The steps to do that is a little bit technical; hence, the following steps were adopted.

Step 1: Compute the first row:

	fx =-0.2429*A1^4+4.7329*A1^3 - 28.396*A1^2 + 64.144*A1 - 26.293							
A	B	C	D	E	F	G	H	I
15.32	14.8	-2070.6072	-1607.7272	12.9938772				

$$x_{n+1} = x_n - \frac{(x_n - x_{n-1})\left(-0.2429x_n^4 + 4.7329x_n^3 - 28.396x_n^2 + 64.144x_n - 26.293\right)}{\left(0.2429x_n^4 + 4.7329x_n^3 - 28.396x_n^2 + 64.144x_n - 26.293\right) - \left(0.2429x_{n-1}^4 + 4.7329x_{n-1}^3 - 28.396x_{n-1}^2 + 64.144x_{n-1} - 26.293\right)}$$

	fx =-0.2429*B1^4+4.7329*B1^3 - 28.396*B1^2 + 64.144*B1 - 26.293						
A	B	C	D	E	F	G	H
15.32	14.8	-2070.6072	-1607.7272	12.9938772			

	fx =A1-(((A1-B1)*C1)/(C1-D1))			
A	B	C	D	E
15.32	14.8	-2070.6072	-1607.7272	12.9938772

Step 2: Compute the second row:

| | fx | =-0.2429*E1^4+4.7329*E1^3 - 28.396*E1^2 + 64.144*E1 - 26.293 |

A	B	C	D	E	F	G	H
15.32	14.8	-2070.6072	-1607.7272	12.9938772			
		-528.12866	-2070.6072	12.1974368			

| | fx | =-0.2429*A1^4+4.7329*A1^3 - 28.396*A1^2 + 64.144*A1 - 26.293 |

A	B	C	D	E	F	G	H
15.32	14.8	-2070.6072	-1607.7272	12.9938772			
		-528.12866	-2070.6072	12.1974368			

| | fx | =E1-(((E1-A1)*C2)/(C2-D2)) |

A	B	C	D	E
15.32	14.8	-2070.6072	-1607.7272	12.9938772
		-528.12866	-2070.6072	12.1974368

Step 3: Compute the third row:

| | fx | =-0.2429*E2^4+4.7329*E2^3 - 28.396*E2^2 + 64.144*E2 - 26.293 |

A	B	C	D	E	F	G	H	I
15.32	14.8	-2070.6072	-1607.7272	12.9938772				
		-528.12866	-2070.6072	12.1974368				
		-256.30114	-528.12866	11.446488				

fx =-0.2429*E1^4+4.7329*E1^3 - 28.396*E1^2 + 64.144*E1 - 26.293

A	B	C	D	E	F	G	H
15.32	14.8	-2070.6072	-1607.7272	12.9938772			
		-528.12866	-2070.6072	12.1974368			
		-256.30114	-528.12866	11.446488			

fx =E2-(((E2-E1)*C3)/(C3-D3))

A	B	C	D	E
15.32	14.8	-2070.6072	-1607.7272	12.9938772
		-528.12866	-2070.6072	12.1974368
		-256.30114	-528.12866	11.446488

Step 4: Drag the third row to convergence level:

A	B	C	D	E
15.32	14.8	-2070.6072	-1607.7272	12.9938772
		-528.12866	-2070.6072	12.1974368
		-256.30114	-528.12866	11.446488
		-84.252855	-256.30114	11.0787448
		-24.425624	-84.252855	10.9286065
		-3.9894536	-24.425624	10.8992972
		-0.2524734	-3.9894536	10.897317
		-0.0029078	-0.2524734	10.897294
		-2.161E-06	-0.0029078	10.897294
		-1.852E-11	-2.161E-06	10.897294
		-2.203E-13	-1.852E-11	10.897294

If this is done, the root of the graph is 10.897294 and the convergence graph is shown below. The convergence showed that at the root of the dataset was obtained at the eighth iteration.

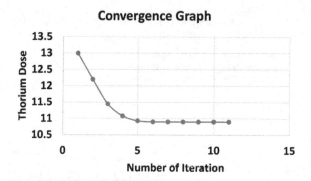

Convergence Graph

1.2.2.1.3 Using the bisection method

Using Eq. (5.1)

$$x_{n+1} = x_{n-1} + \frac{x_n - x_{n-1}}{2}$$

Unlike the previous method, this method is applied directly on the dataset. In this case, the Microsoft Excel Package can be used to compute the solutions. The steps to do that is tricky and burdensome; hence, the following steps were adopted:

Case 1

Step 1: Compute the first row for data 1 and 2.

1   ▲▼   × ✓ f_x =A1+((A2-A1)/2)

A	B	C	D
13.45	19.155	19.3	17.675

Step 2: Compute the second row for data 2 and first result.

▲▼   × ✓ f_x =A2+((B1-A2)/2)

A	B	C	D
13.45	19.155	19.3	17.675
24.86	22.0075	16.52	19.6425

Step 3: Compute the third row for first and second result.

| | X | ✓ | *fx* | =B1+((B2-B1)/2) |

A	B	C	D
13.45	19.155	19.3	17.675
24.86	22.0075	16.52	19.6425
13.74	20.58125	17.91	18.65875

Step 4: Drag the third row to see the convergence.

	A	B
1	13.45	19.155
2	24.86	22.0075
3	13.74	20.58125
4	21.61	21.294375
5	21.84	20.937813
6	44.29	21.116094
7		21.026953
8		21.071523
9		21.049238
10		21.060381
11		21.05481
12		21.057595
13		21.056202
14		21.056899
15		21.056551
16		21.056725
17		21.056638
18		21.056681
19		21.056659
20		21.05667
21		21.056665
22		21.056668
23		21.056666
24		21.056667
25		21.056667
26		21.056667
27		21.056667
28		21.056667
29		21.056667
30		21.056667
31		21.056667

Case 2

Step 1: Compute the first row for data 2 and 3.

1	⬍	✕ ✓	fx	=A2+((A3-A2)/2)

A	B	C	D
13.45	19.155	19.3	17.675

Step 2: Compute the second row for data 3 and first result.

!	⬍	✕ ✓	fx	=A3+((C1-A3)/2)

A	B	C	D	
13.45	19.155	19.3	17.675	
24.86	22.0075	16.52	19.6425	2

Step 3: Compute the third row for first and second result.

}	⬍	✕ ✓	fx	=C1+((C2-C1)/2)

A	B	C	D	
13.45	19.155	19.3	17.675	
24.86	22.0075	16.52	19.6425	
13.74	20.58125	17.91	18.65875	2

Step 4: Drag the third row to see the convergence.

	A	B	C
1	13.4	Name Box 55	19.3
2	24.86	22.0075	16.52
3	13.74	20.58125	17.91
4	21.61	21.294375	17.215
5	21.84	20.937813	17.5625
6	44.29	21.116094	17.38875
7		21.026953	17.475625
8		21.071523	17.432188
9		21.049238	17.453906
10		21.060381	17.443047
11		21.05481	17.448477
12		21.057595	17.445762
13		21.056202	17.447119
14		21.056899	17.44644
15		21.056551	17.44678
16		21.056725	17.44661
17		21.056638	17.446695
18		21.056681	17.446653
19		21.056659	17.446674
20		21.05667	17.446663
21		21.056665	17.446668
22		21.056668	17.446666
23		21.056666	17.446667
24		21.056667	17.446666
25		21.056667	17.446667
26		21.056667	17.446667
27		21.056667	17.446667
28		21.056667	17.446667
29		21.056667	17.446667
30		21.056667	17.446667
31		21.056667	17.446667

Case 3

Step 1: Compute the first row for data 3 and 4.

1	▲▼	✕ ✓	fx =A3+((A4-A3)/2)	
A	B	C	D	
13.45	19.155	19.3	17.675	

Step 2: Compute the second row for data 4 and first result.

)2	▲▼	✕ ✓	fx =A4+((D1-A4)/2)	
A	B	C	D	
13.45	19.155	19.3	17.675	
24.86	22.0075	16.52	19.6425	2

Step 3: Compute the third row for first and second result.

3	▲▼	✕ ✓	fx =D1+((D2-D1)/2)	
A	B	C	D	
13.45	19.155	19.3	17.675	2
24.86	22.0075	16.52	19.6425	21
13.74	20.58125	17.91	18.65875	21.

Step 4: Drag the third row to see the convergence.

	A	B	C	D
1	13.45	19.155	19.3	17.675
2	24.86	22.0075	16.52	19.6425
3	13.74	20.58125	17.91	18.65875
4	21.61	21.294375	17.215	19.150625
5	21.84	20.937813	17.5625	18.904688
6	44.29	21.116094	17.38875	19.027656
7		21.026953	17.475625	18.966172
8		21.071523	17.432188	18.996914
9		21.049238	17.453906	18.981543
10		21.060381	17.443047	18.989229
11		21.05481	17.448477	18.985386
12		21.057595	17.445762	18.987307
13		21.056202	17.447119	18.986346
14		21.056899	17.44644	18.986827
15		21.056551	17.44678	18.986587
16		21.056725	17.44661	18.986707
17		21.056638	17.446695	18.986647
18		21.056681	17.446653	18.986677
19		21.056659	17.446674	18.986662
20		21.05667	17.446663	18.986669
21		21.056665	17.446668	18.986665
22		21.056668	17.446666	18.986667
23		21.056666	17.446667	18.986666
24		21.056667	17.446666	18.986667
25		21.056667	17.446667	18.986667
26		21.056667	17.446667	18.986667
27		21.056667	17.446667	18.986667
28		21.056667	17.446667	18.986667
29		21.056667	17.446667	18.986667
30		21.056667	17.446667	18.986667
31		21.056667	17.446667	18.986667

Case 4

Step 1: Compute the first row for data 4 and 5.

		fx	=A4+((A5-A4)/2)	
A	B	C	D	E
13.45	19.155	19.3	17.675	21.725

Step 2: Compute the second row for data 5 and first result.

		fx	=A5+((E1-A5)/2)	
A	B	C	D	E
13.45	19.155	19.3	17.675	21.725
24.86	22.0075	16.52	19.6425	21.7825

Step 3: Compute the third row for first and second result.

		fx	=E1+((E2-E1)/2)	
A	B	C	D	E
13.45	19.155	19.3	17.675	21.725
24.86	22.0075	16.52	19.6425	21.7825
13.74	20.58125	17.91	18.65875	21.75375

Step 4: Drag the third row to see the convergence.

	A	B	C	D	E
1	13.45	19.155	19.3	17.675	21.725
2	24.86	22.0075	16.52	19.6425	21.7825
3	13.74	20.58125	17.91	18.65875	21.75375
4	21.61	21.294375	17.215	19.150625	21.768125
5	21.84	20.937813	17.5625	18.904688	21.760938
6	44.29	21.116094	17.38875	19.027656	21.764531
7		21.026953	17.475625	18.966172	21.762734
8		21.071523	17.432188	18.996914	21.763633
9		21.049238	17.453906	18.981543	21.763184
10		21.060381	17.443047	18.989229	21.763408
11		21.05481	17.448477	18.985386	21.763296
12		21.057595	17.445762	18.987307	21.763352
13		21.056202	17.447119	18.986346	21.763324
14		21.056899	17.44644	18.986827	21.763338
15		21.056551	17.44678	18.986587	21.763331
16		21.056725	17.44661	18.986707	21.763335
17		21.056638	17.446695	18.986647	21.763333
18		21.056681	17.446653	18.986677	21.763334
19		21.056659	17.446674	18.986662	21.763333
20		21.05667	17.446663	18.986669	21.763333
21		21.056665	17.446668	18.986665	21.763333
22		21.056668	17.446666	18.986667	21.763333
23		21.056666	17.446667	18.986666	21.763333
24		21.056667	17.446666	18.986667	21.763333
25		21.056667	17.446667	18.986667	21.763333
26		21.056667	17.446667	18.986667	21.763333
27		21.056667	17.446667	18.986667	21.763333
28		21.056667	17.446667	18.986667	21.763333
29		21.056667	17.446667	18.986667	21.763333
30		21.056667	17.446667	18.986667	21.763333
31		21.056667	17.446667	18.986667	21.763333

Case 5

Step 1: Compute the first row for data 5 and 6.

1	⏷	× ✓	fx	=A5+((A6-A5)/2)	
A	B	C	D	E	F
13.45	19.155	19.3	17.675	21.725	33.065
~~24.86~~	~~22.0075~~	~~16.52~~	~~19.6425~~	~~21.7825~~	~~38.6775~~

Step 2: Compute the second row for data 6 and first result.

2	⏷	× ✓	fx	=A6+((F1-A6)/2)	
A	B	C	D	E	F
13.45	19.155	19.3	17.675	21.725	33.065
24.86	22.0075	16.52	19.6425	21.7825	38.6775
~~13.74~~	~~20.58125~~	~~17.91~~	~~18.65875~~	~~21.75375~~	~~35.87125~~

Step 3: Compute the third row for first and second result.

	⏷	× ✓	fx	=F1+((F2-F1)/2)	
A	B	C	D	E	F
13.45	19.155	19.3	17.675	21.725	33.065
24.86	22.0075	16.52	19.6425	21.7825	38.6775
13.74	20.58125	17.91	18.65875	21.75375	35.87125

Step 4: Drag the third row to see the convergence.

	A	B	C	D	E	F
1	13.45	19.155	19.3	17.675	21.725	33.065
2	24.86	22.0075	16.52	19.6425	21.7825	38.6775
3	13.74	20.58125	17.91	18.65875	21.75375	35.87125
4	21.61	21.294375	17.215	19.150625	21.768125	37.274375
5	21.84	20.937813	17.5625	18.904688	21.760938	36.572813
6	44.29	21.116094	17.38875	19.027656	21.764531	36.923594
7		21.026953	17.475625	18.966172	21.762734	36.748203
8		21.071523	17.432188	18.996914	21.763633	36.835898
9		21.049238	17.453906	18.981543	21.763184	36.792051
10		21.060381	17.443047	18.989229	21.763408	36.813975
11		21.05481	17.448477	18.985386	21.763296	36.803013
12		21.057595	17.445762	18.987307	21.763352	36.808494
13		21.056202	17.447119	18.986346	21.763324	36.805753
14		21.056899	17.44644	18.986827	21.763338	36.807123
15		21.056551	17.44678	18.986587	21.763331	36.806438
16		21.056725	17.44661	18.986707	21.763335	36.806781
17		21.056638	17.446695	18.986647	21.763333	36.80661
18		21.056681	17.446653	18.986677	21.763334	36.806695
19		21.056659	17.446674	18.986662	21.763333	36.806652
20		21.05667	17.446663	18.986669	21.763333	36.806674
21		21.056665	17.446668	18.986665	21.763333	36.806663
22		21.056668	17.446666	18.986667	21.763333	36.806668
23		21.056666	17.446667	18.986666	21.763333	36.806666
24		21.056667	17.446666	18.986667	21.763333	36.806667
25		21.056667	17.446667	18.986667	21.763333	36.806666
26		21.056667	17.446667	18.986667	21.763333	36.806667
27		21.056667	17.446667	18.986667	21.763333	36.806667
28		21.056667	17.446667	18.986667	21.763333	36.806667
29		21.056667	17.446667	18.986667	21.763333	36.806667
30		21.056667	17.446667	18.986667	21.763333	36.806667
31		21.056667	17.446667	18.986667	21.763333	36.806667

If this is done, the roots of the dataset are 21.056667, 17.446667, 18.986667, 21.763333, and 36.806667, and the convergence graph for each case are shown below. The convergence showed that at the root of the dataset was obtained at the 24th iteration. The clumsiness and error prone nature of the calculation makes it difficult.

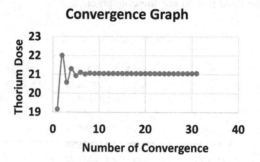

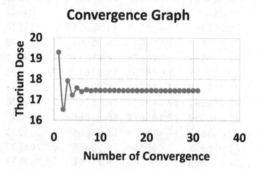

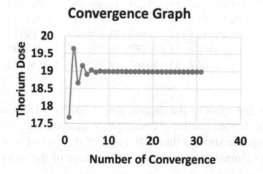

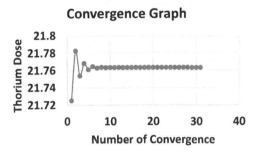

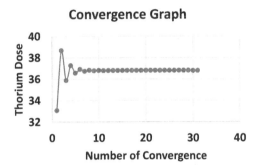

1.3 Computational application of root finding method to data application

This section is very important for two reasons:

1. To equip the environmentalist on working on mathematical theories or models of field work
2. To equip the environmentalist on comparing theory and live dataset to understand where and when outliers appear during field measurement
3. To equip the environmentalist to be able to work on large/big data

In Table 5.1, we determined the equation expressing the Thorium dosage rate in a field dataset. Now imagine how cumbersome it will be if we perform a comparative analysis of the three dose rates: i.e., Thorium, Potassium, and Radon. This is where computational approach becomes handy to solve problems at very short time. The

MATLAB software will be adopted for this illustration. It is advisable that non-MATLAB users can follow through the ideas and apply same to whatsoever software they are using.

1.3.1 Computational application to Newton's method

Let us recap on Table 5.1 where the equation describing the dataset is given for thorium, potassium and radon as:

$$y = -0.2429x^4 + 4.7329x^3 - 28.396x^2 + 64.144x - 26.293 \tag{5.7}$$

$$y = -0.6306x^4 + 7.0225x^3 - 22.634x^2 + 21.184x + 12.038 \tag{5.8}$$

$$y = -0.1808x^4 + 3.2911x^3 - 18.801x^2 + 39.633x - 7.24 \tag{5.9}$$

MATLAB and its users have huge online library of existing work. In other words, you do not have to reinvent the will when most of the job has been solved. For example, this section would adopt the existing root functions written by Tamas (2022). We shall only be interested in writing the code to illustrate the solutions. The MATLAB version used for this computation was 2017b.

We want to solve Eq. (5.7) using the function "newtons_method.m" written by Tamas (2022). The code (i.e., Code5.1) is written below.

```
% Write out the equation you want to solve
   1.  f = @(x) -0.2429*x.^4 + 4.7329*x.^3 - 28.396*x.^2 + 64.144*x - 26.293;
% Write out the differentiation of the above equation
   2.  df = @(x) -0.9716*x.^3+14.1987*x.^2 - 56.792*x + 64.144;
   3.  opts.return_all = true;
%recall that assumption we made earlier was 14.8
   4.  solved_roots = newtons_method(f,df,14.8,opts);
%Initiate the plot canvass
   5.  figure;
%plot with defined parameters
   6.  plot(solved_roots,'-k*','MarkerSize',9,'LineWidth',1.5);
   7.  grid on;
%label the axes
   8.  xlabel('Number of Iteration','Interpreter','latex','FontSize',18);
   9.  ylabel('Thorium Dose','Interpreter','latex','FontSize',18);
```

Take note that your code will not run if you do not download the function "newtons_method.m" and include it in the same folder as your code. The graph is presented below. You will observe that it is the same as the one done using Microsoft Excel. At this juncture, you will permit to say that users should adopt whatever software that is more convenient for them.

Another challenge users may face is having a fair knowledge of mathematics, i.e., ability to differentiate the main equation as presented in code5.1 above. The idea of this section is to assist users who are inconvenient with mathematical procedure. Hence, users in this category can apply code5.2 below.

% Represent the equation as a symbol
   1. **syms** f(x)
% Write out the equation you want to solve
   2. **f(x) = -0.2429*x.^4 + 4.7329*x.^3 - 28.396*x.^2 + 64.144*x - 26.293;**
   3. **df = diff(f);**
   4. **opts.return_all = true;**
%recall that assumption we made earlier was 14.8
   5. **solved_roots = newtons_method(f,df,14.8,opts);**
%Initiate the plot canvass
   6. **figure;**
%plot with defined parameters
   7. **plot(solved_roots,'-k*','MarkerSize',9,'LineWidth',1.5);**
   8. **grid** on;
%label the axes
   9. **xlabel('Number of Iteration','Interpreter','latex','FontSize',18);**
  10. **ylabel('Thorium Dose','Interpreter','latex','FontSize',18);**

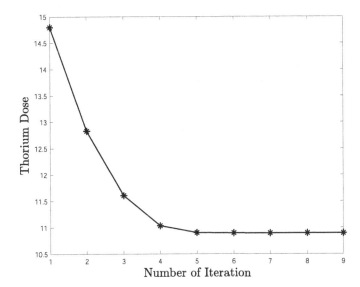

Hence, the convergence graph for potassium (Table 5.1) is:

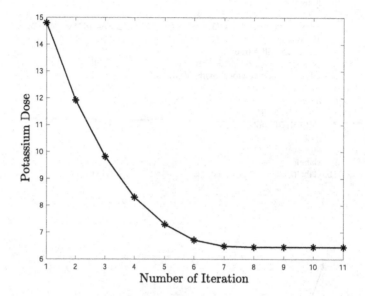

It can be seen that the convergence of potassium had 11 iterations compared to the convergence graph for thorium.

Let us move on to a more complex scenario, i.e., finding the intersect between two functions (recall in this case is the dataset). Finding the intersection of the two functions is equivalent to finding the root of their difference. In doing that, we wrote the code5.3 below

```
1.  clc;clear;
% Represent the equation as a symbol
2.  syms y1(x)
% Write out the equation you want to solve
3.  y1(x) = -0.2429*x.^4 + 4.7329*x.^3 - 28.396*x.^2 + 64.144*x - 26.293;
4.  dy1 = diff(y1);
% Repeat same process for the second equation
5.  syms y2(x)
6.  y2(x) = -0.6306*x.^4 + 7.0225*x.^3 - 22.634*x.^2 + 21.184*x + 12.038;
7.  dy2 = diff(y2);
% defines interval for plotting (Recall the number of location in Table 5.1)
8.  x = 0:0.1:6;
% define the plot option
9.  opts.return_all = true;
%recall that assumption we made earlier was 14.8
10. solved_roots1 = newtons_method(y1,dy1,14.8,opts);
11. solved_roots2 = newtons_method(y2,dy2,14.8,opts);
%Initiate the plot canvass
12. figure;
13. hold on;
%Plot the convergence graph of the two locations
14. plot(solved_roots1,'-k*','MarkerSize',9,'LineWidth',1.5);
15. plot(solved_roots2,'-r*','MarkerSize',9,'LineWidth',1.5);
16. hold off;
17. grid on;
%Label the axes
18. xlabel('Number of Iterations','Interpreter','latex');
19. ylabel('Dose rate','Interpreter','latex');
20. legend('Thorium','Potassium','Interpreter','latex','Location','northeast');
21. set(gca, 'FontSize',18);
%Set the intersection of both equations
22. x_int = newtons_method(@(x) y2(x)-y1(x),@(x) dy2(x)-dy1(x),3.6,opts);
%Initiate the plot canvass for the second plot
23. figure;
24. hold on;
%Plot the graph of the two locations and find the intersect
25. plot(x,y1(x),'LineWidth',1.5);
26. plot(x,y2(x),'LineWidth',1.5);
27. plot(x_int,y1(x_int),'ko','MarkerSize',9,'LineWidth',1.5);
28. hold off;
29. grid on;
%Label the axes
30. xlabel('Location','Interpreter','latex');
31. ylabel('Dose rate','Interpreter','latex');
32. legend('Thorium','Potassium','Intersection','Interpreter','latex','Location','southeast');
33. set(gca, 'FontSize',18);
```

The resulting graphs is shown below. The first graph is the convergence graph for both datasets. The second graph is a root intersect of both datasets.

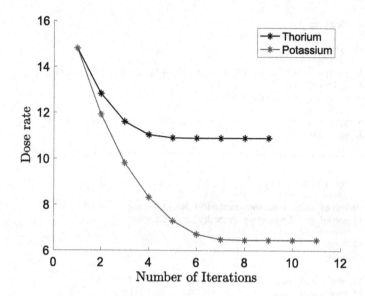

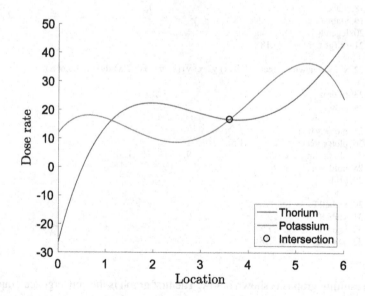

When you take the wrong assumption for the roots of the difference between two functions, you will observe that the intersection circles scatters. When is this process important? Let us assume that an environmentalist wishes to find the source of pollution on a field; this process would enable the researcher to narrow down to the source. This technique is quite reliable than the native knowledge which assumes that the source of pollution would have the highest magnitude of the parameter measured. In reality, the primary source of pollution might had created other secondary sources that have higher concentrations. A good example is investigating an infested farmland.

1.3.2 Computational application to secant method

The secant method had been described with respect to its convergence and accuracy. The computational approach to this method will certainly be less stressful than the manual calculation; hence, the number of iterations will not matter to us but the accuracy. The secant function ("secant_method.m") that was adopted for this section was written by Tamas (2022). Let us recap on Table 5.1 where the equation describing the dataset is given for thorium, potassium, and radon as:

$$y = -0.2429x^4 + 4.7329x^3 - 28.396x^2 + 64.144x - 26.293 \tag{5.7}$$

$$y = -0.6306x^4 + 7.0225x^3 - 22.634x^2 + 21.184x + 12.038 \tag{5.8}$$

$$y = -0.1808x^4 + 3.2911x^3 - 18.801x^2 + 39.633x - 7.24 \tag{5.9}$$

Using Eq. (5.7), the MATLAB code was written as displayed in code5.4 below:

```
% Represent the equation as a symbol
    1. syms f(x)
% Write out the equation you want to solve
    2. f(x) = -0.2429*x.^4 + 4.7329*x.^3 - 28.396*x.^2 + 64.144*x - 26.293;
    3. opts.return_all = true;
%recall that assumption we made earlier was 14.8
    4. solved_roots = secant_method(f,14.8,opts);
%Initiate the plot canvass
    5. figure;
%plot with defined parameters
    6. plot(solved_roots,'-k*','MarkerSize',9,'LineWidth',1.5);
    7. grid on;
%label the axes
    8. xlabel('No of Iteration','Interpreter','latex','FontSize',18);
    9. ylabel('Thorium Dose','Interpreter','latex','FontSize',18);
```

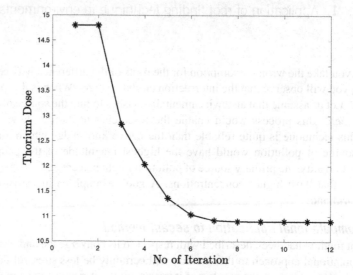

If two datasets are analyzed, the code to get the convergence graph and root intersect of both datasets is displayed in code5.5 below.

```
% Write out the equations you want to solve
    1.  y1 = @(x) -0.2429*x.^4 + 4.7329*x.^3 - 28.396*x.^2 + 64.144*x - 26.293;
    2.  y2 = @(x) -0.6306*x.^4 + 7.0225*x.^3 - 22.634*x.^2 + 21.184*x + 12.038;
% defines interval for plotting (Recall the number of location in Table 5.1)
    3.  x = 0:0.1:6;
% define the plot option
    4.  opts.return_all = true;
%recall that assumption we made earlier was 14.8
    5.  solved_roots1 = secant_method(y1,14.8,opts);
    6.  solved_roots2 = secant_method(y2,14.8,opts);
%Initiate the plot canvass
    7.  figure;
    8.  hold on;
    9.  plot(solved_roots1,'-k*','MarkerSize',9,'LineWidth',1.5);
    10. plot(solved_roots2,'-r*','MarkerSize',9,'LineWidth',1.5);
    11. hold off;
    12. grid on;
%Label the axes
    13. xlabel('Number of Iterations','Interpreter','latex');
    14. ylabel('Dose rate','Interpreter','latex');
    15. legend('Thorium','Potassium','Interpreter','latex','Location','northeast');
    16. set(gca, 'FontSize',18);
%Set the intersection of both equations
    17. x_int = secant_method(@(x) y2(x)-y1(x),3.6);
    18. figure;
    19. hold on;
    20. plot(x,y1(x),'LineWidth',1.5);
    21. plot(x,y2(x),'LineWidth',1.5);
    22. plot(x_int,y1(x_int),'ko','MarkerSize',9,'LineWidth',1.5);
    23. hold off;
    24. grid on;
%Label the axes
    25. xlabel('Location','Interpreter','latex');
    26. ylabel('Dose rate','Interpreter','latex');
    27. legend('Thorium','Potassium','Intersection','Interpreter','latex','Location','southeast');
    28. set(gca, 'FontSize',18);
```

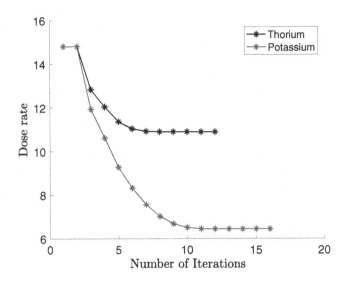

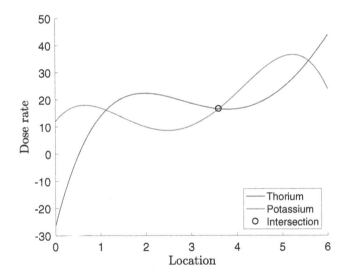

When you compare the results of the convergence graph and root intersect of both datasets using the Newton's and Secant's method, it is clear that the accuracy is the same for both. The difference between both methods is higher number of iterations. For example, the potassium dose had 11 iterations in the Newton's method while it has 16 iterations in the Secant's method. Also, the pattern of the convergence graph is a little different from the graph displayed in the Newton's method. Most persons prefer the secant method using computational tool because of the ease to write its code which is relative in the researcher task and perception.

1.3.3 Computational application to bisection method

Recall the bisection method is referred as least accurate among the root techniques with highest number of iterations. However, computational approach makes any of the methods easier as the number of iterations will not matter to us but the accuracy. The bisection function ("bisection_method.m") that was adopted for this section was written by Tamas (2022). Like we did for the previous subsections, we recap on Table 5.1 where the equation describing the dataset is given for thorium, potassium and radon as:

$$y = -0.2429x^4 + 4.7329x^3 - 28.396x^2 + 64.144x - 26.293 \tag{5.7}$$

$$y = -0.6306x^4 + 7.0225x^3 - 22.634x^2 + 21.184x + 12.038 \tag{5.8}$$

$$y = -0.1808x^4 + 3.2911x^3 - 18.801x^2 + 39.633x - 7.24 \tag{5.9}$$

Using Eq. (5.7), the MATLAB code was written as displayed in code5.6 below:

```
% Write out the equation you want to solve
  1.  f = @(x) -0.2429*x.^4 + 4.7329*x.^3 - 28.396*x.^2 + 64.144*x - 26.293;
  2.  opts.return_all = true;
%recall that assumption we made earlier was 21.675
  3.  solved_roots = bisection_method (f,0,21.675, opts);
%Initiate the plot canvass
  4.  figure;
%plot with defined parameters
  5.  plot(solved_roots,'-k*','MarkerSize',9,'LineWidth',1.5);
  6.  grid on;
%label the axes
  7.  xlabel('No of Iteration','Interpreter','latex','FontSize',18);
  8.  ylabel('Thorium Dose','Interpreter','latex','FontSize',18);
```

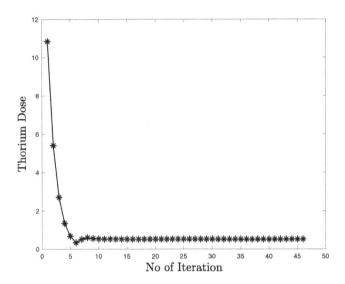

From the above graph it is clear why professionals do not choose the bisection method. The accuracy is relative to the assumption and are not consistent. For example, if the assumption is increased by the value 1, i.e., 22.675, the graph will take another pattern as presented below. More so, the number of iterations thrice any of Newton's or secant method.

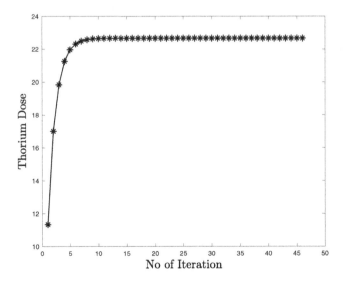

It now clear why bisection is not considered in environmental computation. However, some persons still use it.

If two datasets are analyzed, the code to get the convergence graph and root intersect of both datasets is displayed in code5.7 below.

```matlab
% Write out the equations you want to solve
1.   y1 = @(x) -0.2429*x.^4 + 4.7329*x.^3 - 28.396*x.^2 + 64.144*x - 26.293;
2.   y2 = @(x) -0.6306*x.^4 + 7.0225*x.^3 - 22.634*x.^2 + 21.184*x + 12.038;
% defines interval for plotting (Recall the number of location in Table 5.1)
3.   x = 0:0.1:6;
% define the plot option
4.   opts.return_all = true;
%recall that assumption we made earlier was 22.675
5.   solved_roots1 = bisection_method(y1,0,22.675,opts);
6.   solved_roots2 = bisection_method(y2,0,22.675,opts);
%Initiate the plot canvass
7.   figure;
8.   hold on;
9.   plot(solved_roots1,'-k*','MarkerSize',9,'LineWidth',1.5);
10.  plot(solved_roots2,'-r*','MarkerSize',9,'LineWidth',1.5);
11.  hold off;
12.  grid on;
%Label the axes
13.  xlabel('Number of Iterations','Interpreter','latex');
14.  ylabel('Dose rate','Interpreter','latex');
15.  legend('Thorium','Potassium','Interpreter','latex','Location','northeast');
16.  set(gca, 'FontSize',18);
%Set the intersection of both equations
17.  x_int = bisection_method(@(x) y2(x)-y1(x),3.5,3.6);
18.  figure;
19.  hold on;
20.  plot(x,y1(x),'LineWidth',1.5);
21.  plot(x,y2(x),'LineWidth',1.5);
22.  plot(x_int,y1(x_int),'ko','MarkerSize',9,'LineWidth',1.5);
23.  hold off;
24.  grid on;
%Label the axes
25.  xlabel('Location','Interpreter','latex');
26.  ylabel('Dose rate','Interpreter','latex');
27.  legend('Thorium','Potassium','Intersection','Interpreter','latex','Location','southeast');
28.  set(gca, 'FontSize',18);
```

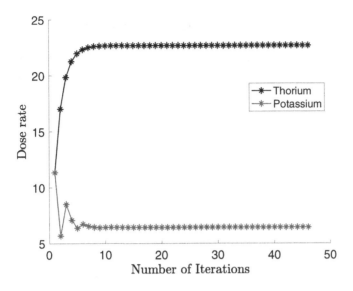

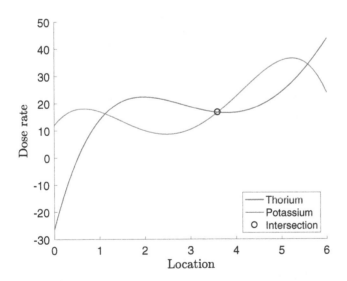

It is observed that only the root intersect graph looks like other methods discussed earlier. The convergence graph is significantly at variance to the convergence graph presented in the Newton's and Secant's method. Interestingly, if you raise the assumption to higher magnitudes, the patterns remain the same. However, the pattern changes at lower magnitudes, i.e., <1, as presented below.

Another common use of bisection method is when the researcher is interested in seeing the procedural changes in the iterations. This type of analysis is mostly presented in a table. Code5.8 is a representation of how this type of research is performed.

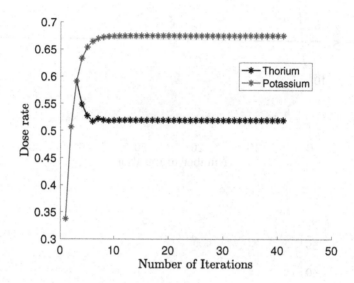

1.3.3.1 Code5.8

```
% Define the table parameters
  1.  a = 13.45; % value of lowest value in thorium of Table 5.1
  2.  b = 44.29; % value of highest value in thorium of Table 5.1
  3.  E = 1e-6; % proposed error values
  4.  N = ceil(log((b - a) / E) / log(2)); % CEIL() !
% Write out the equations you want to solve
  5.  f = @(x) -0.2429*x.^4 + 4.7329*x.^3 - 28.396*x.^2 + 64.144*x - 26.293;
%create the table
  6.  T = cell2table(cell(0,5),'VariableNames', {'N', 'a_n', 'b_n', 'c_n', 'p_n'});
  7.  for i = 1:N
  8.  c = (a + b) / 2; % The bisection of 'a' and 'b'
  9.  T(i, :) = {i, a, b, c, f(c)};  % Fill in data in existing table
  10. if f(c) * f(a) > 0
  11. a = c;
  12. else
  13. b = c;
  14. end
  15. end
  16. disp(T)
```

N	a_n	b_n	c_n	p_n
1	13.45	44.29	28.87	−76695
2	28.87	44.29	36.58	−2.39E+05
3	36.58	44.29	40.435	−3.80E+05
4	40.435	44.29	42.362	−4.71E+05
5	42.362	44.29	43.326	−5.22E+05
6	43.326	44.29	43.808	−5.48E+05
7	43.808	44.29	44.049	−5.62E+05
8	44.049	44.29	44.17	−5.69E+05
9	44.17	44.29	44.23	−5.73E+05
10	44.23	44.29	44.26	−5.75E+05
11	44.26	44.29	44.275	−5.75E+05
12	44.275	44.29	44.282	−5.76E+05
13	44.282	44.29	44.286	−5.76E+05
14	44.286	44.29	44.288	−5.76E+05
15	44.288	44.29	44.289	−5.76E+05
16	44.289	44.29	44.29	−5.76E+05
17	44.29	44.29	44.29	−5.76E+05
18	44.29	44.29	44.29	−5.76E+05
19	44.29	44.29	44.29	−5.76E+05
20	44.29	44.29	44.29	−5.76E+05
21	44.29	44.29	44.29	−5.76E+05
22	44.29	44.29	44.29	−5.76E+05
23	44.29	44.29	44.29	−5.76E+05
24	44.29	44.29	44.29	−5.76E+05
25	44.29	44.29	44.29	−5.76E+05

One insight that this approach gives is when there is a missing value in the measurement that must be filled and the research wants to fill-it up theoretical. Missing values are common with most remote sensing dataset. It occurs when no data value is stored for the variable in an observation. Missing value something may have significant effect on analysis and by extension the conclusions or decision. This type of approximations becomes very necessary provided the missing values are not much. In a case where there are many missing values, it is advisable for the environmental researcher to discard the dataset and repeat the measurement. In the case of remote sensing dataset, it is advisable that the researcher use another satellite station.

Reference

Tamas, K., 2022. Newton's Method (Newtons_method). https://github.com/tamaskis/newtons_method-MATLAB/releases/tag/v5.5.0. GitHub. Retrieved January 15, 2022.

Numerical differential analysis in environmental research

1. Introduction

Solving differential equations is like driving a car where the decision to speedup or slowdown is at driver's prerogative. In this case, the solver may decide to use linear, separable, slope fields, directional fields, integrable factor, or exact differential equations or differential equations that could be solved with a set of very specific substitutions. What drives the process of solving differential equation is: the complex nature of the equation; specific interest in the behavior of the solution; and getting general trends in solutions and for long-term behavior of solutions. However, it is preferred that the differential equation is solved for an explicit solution that is informative with inductive interpretation. The first-order equations could be divided into the linear equation, separable equation, nonlinear equation, exact equation, homogeneous equation, Bernoulli equation, and nonhomogeneous equations. Most models in environmental studies are mostly in differential format. The differential format can be:

1. Simple first-order ordinary differential equation;
2. Simple second-order ordinary differential equation;
3. Simple third-order ordinary differential equation;
4. Group of first-order ordinary differential equation;
5. Group of second-order ordinary differential equation;
6. Group of third-order ordinary differential equation;
7. Complex first-order ordinary differential equation;
8. Complex second-order ordinary differential equation; and
9. Complex third-order ordinary differential equation.

In this section, an in-depth explanation of each category of differential equations are considered so that reader can identify the type of differential pattern in a model that chooses an appropriate solving technique.

A typical example of this type of differential equation is the first-order homogeneous linear differential equation which has the form

$$\frac{dy}{dx} + f(x)y = 0 \qquad (6.1)$$

Numerical Methods in Environmental Data Analysis. https://doi.org/10.1016/B978-0-12-818971-9.00007-7

The general solution of this type of equation is given as:

$$\frac{dy}{dx} = -f(x)y$$

$$\frac{1}{y}dy = -f(x)dx$$

$$\int \frac{1}{y}dy = \int -f(x)dx$$

$$In|y| = f(x) + c$$

$$y = \pm Ae^{f(x)+c}$$

$$y = Ae^{f(x)}$$

Hence, if an equation of this sort is seen, the general solution brings us closer to the answer. The solution of A could be obtained using the initial value problem. For example,

$$\frac{dy}{dt} - 5ty = 0 \qquad (6.2)$$

for $y(0) = 6$

From the general solution, the

$$y = Ae^{f(t)}$$

$$f(t) = \int 5tdt = 2.5t^2$$

$$y = Ae^{2.5t^2}$$

Hence, applying the IVP

$$6 = Ae^{2.5*0^2}$$

$$A = 6$$

Therefore, $y = 6e^{2.5t^2}$.

In practical terms, solving differential equation numerically are not invoke to core fields of environmental science. Hence, what makes sense to most professionals is the computational application of this numerical analysis as it is fast, less cumbersome, less stressful to manipulate, more accurate, and more flexible to use. In the formulation of environmental problem especially investigating growth rate or infection rate in a laboratory scale, a simple second-order equation are mostly used. However, there may be few exceptions that can be complex in nature. In this section, the numerical solutions of a simple second-order ordinary differential equation is explained below. Generally, there is the assignment of boundary conditions at two separate points for this type of differentiation. This type of boundary condition is referred to as a two-point boundary value problem. The boundary problem used to solve Eq. (6.3) is known as the initial value problem (IVP) where the boundary condition is represented by a single point. A typical example is the solved example in Eq. (6.4). Two-point boundary value problems are more complex than the IVP because it requires more complex process.

The numerical methods for solving ordinary differential equation include but not limited to Euler method, Improved Euler method, RK method, Mid-point method, Predictor Corrector method etc.

1.1 Euler method

Euler method is a common method for solving first-order numerical ordinary differential equations (ODEs) with a given initial value. It is also referred to as forward Euler method or simplest Runge–Kutta method (i.e., First-Order Runge–Kutta) and the simplest method for numerical integration of ordinary differential equations. The procedural steps to solving ODE via Euler method is prone to local error (error per step) which is proportional to the square of the step size.

The theory of the Euler's method starts with the assumption that $y(x_1)$ denotes exact solution and y_1 the computed solution at $x_i = x_0 + h$ such that

$$y(x_1) = y(x_0 + h) \tag{6.3}$$

The solution can be derived from the Taylor's series given as:

$$y(x_0 + h) = y(x_0) + hy'(x_0) + \frac{h^2}{2}y''(x_0) + \frac{h^3}{3}y'''(x_0) + \dots \tag{6.4}$$

To apply the Taylor's series here means that Eq. (6.4) will be truncated after second term

$$y(x_1) = y(x_0) + hy'(x_0) \tag{6.5}$$

to give the Euler's formula so that,

$$y_1 = y_0 + hf(x_0, y_0) \tag{6.6}$$

$f(x_0, y_0)$ is a known function and the values in the initial condition are also known numbers. Function is assumed to be continuous so that a unique solution to the initial value problem (IVP) will be obtained. The error in the formula is the third term of Eq. (6.4) i.e., $\frac{h^2}{2}y''(\xi)$ where $x_0 \le \xi \le x_1$.

When $x = x_{n+1}$, then the Euler's formula becomes,

$$y_{n+1} = y_n + hf(x_n, y_n) \tag{6.7}$$

Hence, the error in the formula would be $\frac{h^2}{2}y''(\xi), x_n \le \xi \le x_{n+1}$. The error in the Euler's method depends on the magnitude of the step size. Euler's method uses the idea of local linearity or linear approximation; hence, the disadvantage of the Euler's method is the tendency to sometimes fail or cumbersome to solve manually.

A typical application of Euler's is presented below to solve the equation:

$$\frac{dy}{dx} = \frac{y \ln y}{2x}$$

where the initial condition is $y(2) = e$.

From the initial condition, $x_0 = 2$, $y_0 = e$, the step size of $h = 0.1$, where $e = 2.7182818$.

Applying Eq. (6.7),

$$y_1 = y_0 + hf(x_0, y_0)$$

$$y(2.1) = e + 0.1 \frac{eIne}{2 \times 2.1} \approx 2.7830027952$$

$$y_2 = y_1 + hf(x_1, y_1)$$

$$y(2.2) = 2.7830027952 + 0.1 \frac{eIne}{2 \times 2.2} \approx 2.8462528587$$

$$y_3 = y_2 + hf(x_2, y_2)$$

$$y(2.3) = 2.8462528587 + 0.1 \frac{eIne}{2 \times 2.3} \approx 2.9081279208$$

$$y_4 = y_3 + hf(x_3, y_3)$$

$$y(2.4) = 2.9081279208 + 0.1 \frac{eIne}{2 \times 2.4} \approx 2.9687139192$$

$$y_5 = y_4 + hf(x_4, y_4)$$

$$y(2.5) = 2.9687139192 + 0.1 \frac{eIne}{2 \times 2.5} \approx 3.0280881976$$

$$y_6 = y_5 + hf(x_5, y_5)$$

$$y(2.6) = 3.0280881976 + 0.1 \frac{eIne}{2 \times 2.6} \approx 3.0863206629$$

$$y_7 = y_6 + hf(x_6, y_6)$$

$$y(2.7) = 3.0863206629 + 0.1 \frac{eIne}{2 \times 2.7} \approx 3.1434747493$$

$$y_8 = y_7 + hf(x_7, y_7)$$

$$y(2.8) = 3.1434747493 + 0.1 \frac{eIne}{2 \times 2.8} \approx 3.199608227$$

The solution goes on according to the intent of the solver. This process is however very easy when it comes to computation. This process is expressed in succeeding sections.

1.2 Improved Euler method

The improved Euler method is quite similar to the Euler's method. In the improved Euler's method, the approximate solution is improved by using the average of the values at the initially given point and the new point, hence reducing the errors in the Euler's method. We obtain Eqs. (6.8)–(6.10) as:

$$q_1 = f(x_n, y_n) \tag{6.8}$$

$$q_2 = f(x_n + h, y_n + hq_1) \tag{6.9}$$

The average of q_1 and q_2 is used in the Euler's method:

$$y_{n+1} = f(x_n + h, y_n + h(q_1 + q_2)/2) \tag{6.10}$$

The calculation of the improved Euler was demonstrated using the problem solved earlier, i.e.,

$$\frac{dy}{dx} = \frac{yIny}{2x}$$

where the initial condition is $y(2) = e$.

From the initial condition, $x_0 = 2$, $y_0 = e$, the step size of $h = 0.1$, where $e = 2.7182818$.

$$q_1 = f(x_n, y_n)$$

$$q_1 = f(2, e) = \frac{eIne}{2 \times 2} = 0.6795704429$$

$$q_2 = f(x_n + h, y_n + hq_1)$$

$$q_2 = f(2 + 0.1, e + 0.1 * 0.6795704429) = f(2.1, 2.7862388443)$$

$$q_2 = \frac{2.7862388443In(2.7862388443)}{2 \times 2.1} - 0.6797710311$$

$$y_{n+1} = f(x_n + h, y_n + h(q_1 + q_2)/2)$$

$$y_{n+1} = f\left(2 + 0.1, e + 0.1\left(\frac{0.6795704429 + 0.6797710311}{2}\right)\right)$$

$$f(x_n, y_n) = f(2.1, 2.7862488737) = \frac{2.7862488737In(2.7862488737)}{2 \times 2.1} = 0.679775866$$

$$y_1 = 2.7182818 + 0.1 \times 0.679775866 = 2.7862593866$$

From the first term, it is clear that there is a significant difference between the Euler and the improved Euler. Due to the complexity of the improved Euler, the computational solution using the improved Euler is recommended. The computational process will be adequately explained in succeeding sections in the chapter.

1.3 Runge–Kutta method

Runge–Kutta (RK) methods are referred to as a family of implicit and explicit iterative methods. Like the improved Euler method, RK procedure include the Euler method. There are different types of RK method (explicit, embedded, and implicit). The explicit methods include Forward Euler, Explicit midpoint method, Heun's method, Ralston's method, generic second-order method, Kutta's third-order method, generic third-order method, Heun's third-order method, Ralston's third-order method, third-order strong stability preserving Runge–Kutta (SSPRK3), classic fourth-order method, Ralston's fourth-order method, and 3/8-rule fourth-order method. The embedded methods include Heun–Euler, Fehlberg RK1(2), Bogacki–Shampine, Fehlberg, Cash–Karp, and Dormand–Prince. The implicit methods include the backward Euler, implicit midpoint, Crank–Nicolson method, Gauss–Legendre methods, diagonally implicit Runge–Kutta methods, Lobatto methods, and Radau methods. The most popular RK method is the RK4. It is derived from the initial value problem as the Euler's method where:

$$\frac{dy}{dx} = f(x, y), \quad y(x_0) = y_0 \tag{6.11}$$

and the step size is assumed to be greater than zero at all times. The unknown function y is written as:

$$y_{n+1} = y_n + \frac{1}{6}h(q_1 + 2q_2 + 2q_3 + q_4) \tag{6.12}$$

$$x_{n+1} = x_n + h \tag{6.13}$$

For $n > 0$,

$$q_1 = f(x_n, y_n) \tag{6.14}$$

$$q_2 = f\left(x_n + \frac{h}{2}, y_n + h\frac{q_1}{2}\right) \tag{6.15}$$

$$q_3 = f\left(x_n + \frac{h}{2}, y_n + h\frac{q_2}{2}\right) \tag{6.16}$$

$$q_4 = f(x_n + h, y_n + hq_3) \tag{6.17}$$

The difference between the Euler and RK method is that Euler's method makes use of the information on the slope or the derivative of y at the given time step to extrapolate the solution to the next time-step. On the other hand, Runge–Kutta methods make use of the information on the "slope" at more than one point to extrapolate the solution to the future time step.

The calculation of the RK method was demonstrated using the problem solved earlier, i.e.,

$$\frac{dy}{dx} = \frac{y \ln y}{2x}$$

where the initial condition is $y(2) = e$.

From the initial condition, $x_0 = 2$, $y_0 = e$, the step size of $h = 0.1$, where $e = 2.7182818$; hence, we compute for q_1, q_2, q_3, and q_4

$$q_1 = f(x_0, y_0)$$

$$q_1 = f(2, e) = \frac{e \ln e}{2 \times 2} = 0.6795704429$$

$$q_2 = f\left(x_0 + \frac{h}{2}, y_0 + h\frac{q_1}{2}\right)$$

$$q_2 = f(2.05, 2.7522603221) = \frac{2.7522603221 In(2.7522603221)}{2 \times 2.05} = 0.6796220248$$

$$q_3 = f\left(x_0 + \frac{h}{2}, y_0 + h\frac{q_2}{2}\right)$$

$$q_3 = f(2.05, 2.7522629012) = \frac{2.7522629012 In(2.7522629012)}{2 \times 2.05} = 0.6796232907$$

$$q_4 = f(x_n + h, y_n + hq_3)$$

$$q_4 = f(2.1, 2.7862441291) = \frac{2.7862441291 In(2.7862441291)}{2 \times 2.1} = 0.6797735788$$

$$y_1 = y_0 + \frac{1}{6}h(q_1 + 2q_2 + 2q_3 + q_4)$$

$$y_1 = 2.7182818 + \frac{1}{6}0.1(0.6795704429 + 2(0.6796220248) + 2(0.6796232907) + (0.6797735788))$$

$$y_1 = 2.7862429923$$

This iterative process can be extended to 10 terms, i.e , $n = 10$. The table below is a comparison between the first value of y, i.e., y_1, for the Euler, improved Euler and RK.

Term	Euler	Improved Euler	Runge–Kutta
y_1	2.7830027952	2.7862593866	2.7862429923

The second-order RK is estimated using the IVP presented in Eq. (6.11)
For $n > 0$,

$$q_1 = hf(x_n, y_n) \tag{6.18}$$

$$q_2 = hf\left(x_n + \frac{h}{2}, y_n + h\frac{q_1}{2}\right) \tag{6.19}$$

$$y_{n+1} = y_n + q_2 \tag{6.20}$$

$$x_{n+1} = x_n + h \tag{6.21}$$

The third-order RK is estimated using the IVP presented in Eq. (6.11):
For $n > 0$,

$$x_{n+1} = x_n + h \tag{6.22}$$

$$q_1 = hf(x_n, y_n) \tag{6.23}$$

$$q_2 = hf\left(x_n + \frac{h}{2}, y_n + h\frac{q_1}{2}\right) \tag{6.24}$$

$$q_3 = f(x_n + h, y_n + 2q_2 - q_1) \tag{6.25}$$

$$y_{n+1} = y_n + \frac{1}{6}h(q_1 + 4q_2 + 2q_3) \tag{6.26}$$

Runge–Kutta method are known for solving first-order ordinary differential equations. However, the Euler and Runge–Kutta methods can be used to solve higher-order ordinary differential equations or coupled (simultaneous) differential equations as shall be computationally demonstrated in the succeeding sections in this chapter.

1.4 Predictor Corrector method

This method is also known as Heun's method. It works by algorithms designed to integrate ordinary differential equations with the intent of finding an unknown function that satisfies a given differential equation. All such algorithms proceed in two steps: prediction and correction. The "prediction" is the first step which works by fitting a function to the function-values and derivative-values at a preceding set of points to extrapolate this function's value at a subsequent, new point. On the other hand, the "corrector" is the second step which refines the initial approximation by using the predicted value of the function and another method to interpolate that unknown function's value at the same subsequent point. This method can be constructed from the Euler method (an explicit method) and the trapezoidal rule (an implicit method). Generally, the idea behind the predictor-corrector methods is to use a suitable combination of an explicit and an implicit technique to obtain a method with better convergence characteristics.

Considering the construction from the Euler's method, i.e.,

$$\frac{dy}{dx} = f(x, y), \quad y(x_0) = y_0 \tag{6.27}$$

and the step size is assumed to be greater than zero at all times. The predictor step is given:

$$y_{n+1} = y_n + hf(x_n, y_n) \tag{6.28}$$

while the corrector step is given as:

$$y_{n+1} = y_n + \frac{1}{2}hf(x_n, y_n) + f(x_{n+1}, y_{n+1}) \tag{6.29}$$

The predictor step is executed using explicit method to obtain an approximation for the unknown function y. The corrector step is executed using the implicit method, but with the predicted value of y. The corrector step can go as many times to fine-tune the approximation.

Solving higher-order differential equations using explicit method has its own challenges because implicit methods generally have better properties than the explicit ones (for example, the implicit trapezium is second order while the explicit Euler is only first order). The advantage of the Predictor Corrector Method over few numerical methods is determination of the initial value of y using a logical and reliable process. Other methods (e.g., Newton–Raphson) requires an initial guess. This means that if the guess is wrong, the fine-tuning of the approximation is already compromised from the beginning.

The use of the corrector step has its own difficulty as a linear multistep method are prone to errors if the ith time steps are not properly defined. Hence, the i-step implicit method is given as

$$\alpha_i y_{n+i} + \ldots + \alpha_1 y_1 + \alpha_0 y_0 = h\left(\beta_i f_{n+i} + \ldots + \beta_1 f_1 + \beta_0 f_0\right) \qquad (6.30)$$

The calculation of the Predictor Corrector Method is presented using the problem solved earlier, i.e.,

$$\frac{dy}{dx} = \frac{y \ln y}{2x}$$

where the initial condition is $y(2) = e$.

From the initial condition, $x_0 = 2$, $y_0 = e$, the step size of $h = 0.1$, where $e = 2.7182818$

$$f_0 = f(x_0, y_0)$$

$$f_0 = f(2, e) = \frac{e \ln e}{2 \times 2} = 0.6795704429$$

$$y_1^p = y_0 + hf(x_0, y_0)$$

$$y_1^p = 2.7182818 + 0.1 \times 0.6795704429 = 2.7862388443$$

The corrector step is applied using Eq. (6.29)

$$y_1 = y_0 + \frac{1}{2}h(f(x_0, y_0) + f(x_1, y_1))$$

$$y_1 = 2.7182818 + \frac{1}{2} \times 0.1 \times (0.6795704429 + 0.6797710311) = 2.7862488737$$

This process completes the first-time steps of prediction and one correction. For the second time step

$$f_1 = f(h, y_1) = f(0.1, 2.7862488737)$$

$$f_1 = \frac{2.7862488737 \ln(2.7862488737)}{2 \times 0.1} = 14.2752931863$$

The second predictor step becomes

$$y_2^p = y_1 + hf_1$$

$$y_2^p = 2.7862488737 + 0.1 \times 14.2752931863 = 4.2137781923$$

The second corrector step becomes

$$y_2 = y_1 + \frac{1}{2}h\left((f_1 + y_2^p)\right)$$

$$y_2 = 2.7862488737 + \frac{1}{2}0.01(14.2752931863 + 4.2137781923) = 2.8786942306$$

The corrector step is expected to be repeated until the corrected values converge to a certain solution, i.e., depicting that the approximation inherits the properties of the implicit scheme.

1.5 Midpoint method

This method is similar to the Heun's method. Like the predictor-corrector method, it systematically computes the initial value using the explicit method and use the implicit method as the corrector to obtain better convergence of the solution.

Its predictor step is given as:

$$y_{n+\frac{h}{2}} = y_n + \frac{h}{2}f(x_n, y_n) \tag{6.31}$$

Eq. (2.8) is referred as the prediction of the solution value halfway between x_n and x_{n+1}.

The corrector step of the process is given as:

$$y_{n+1} = y_n + f\left(x_n + \frac{h}{2}, y_{n+\frac{h}{2}}\right) \tag{6.32}$$

You may wish to use these equations to solve the problem we had solved in the past.

So far, it has been shown that the step size (h) is a salient factor for the accuracy of the calculation. It is therefore safer for it to starts from 0.05. If h is too large, the approximate solution will be synthesized with high magnitude errors that makes the assumed solution be at variance to the true solution of the differential equation. In same vein, if h is too small, the approximate solution will be synthesized with low magnitude errors makes the process cumbersome with high roundoff errors.

1.6 Application of numerical methods of solving differentiation in environmental research

As discussed in earlier chapter, environmental research is a large field of study that have different concepts in principle. Over the years, differential equations are employed by researchers that care about what is going on in the surroundings (Mungkasi, 2021). For example, pollution is a very large concept based on medium, dynamics, and concept. A good example is studying pollution due to agricultural runoff of a pesticide into a lake. The lake itself is an open system as the previous pollution in the lake is not known, the diffusion rate is not known, and many other challenges. On the other side, the quantity of the pesticide cannot be exactly estimated. Hence, if a model is to be developed, then it will be based on certain assumptions. The known model for pollution of a lake is presented below.

Aguirre and Tully (2019) worked on the formulation of pollution of a lake starting with mixture equation given as:

$$\frac{dM}{dt} = C_{in}(t) * Q_{in}(t) - C_{out}(t) * Q_{out}(t) \qquad (6.33)$$

where C is the concentration of the contaminant, Q is the volumetric flow rate through the lake, M is the mass of the contaminant. Considering the role of the pollutant in influencing the chemical properties of the water, k was introduced to modify Eq. (6.33) as:

$$\frac{dM}{dt} = C_{in}(t) * Q_{in}(t) - C_{out}(t) * Q_{out}(t) - k(t)C(t)V(t) \qquad (6.34)$$

Assumptions to this model includes constant volume of the lake, constant flow rate, constant reaction rate, and a homogenous lake, i.e., well mixed. In the model presented by Prakash and Veerasha (2020), three fractional differential equations describing the Lakes pollution using q-homotopy analysis transform method (q-HATM) was presented. The formulation of the model is based on three different cases of the considered model namely, periodic input model (Eq. 6.33), exponentially decaying input model (Eq. 6.34), and linear input model (Eq. 6.35).

$$\begin{cases} \dfrac{dz_1}{dt} = \dfrac{38}{1180}z_3(t) + 100 - \dfrac{38}{2900}z_1(t) \\[2mm] \dfrac{dz_2}{dt} = \dfrac{8}{2900}z_1(t) - \dfrac{18}{850}z_2(t) \\[2mm] \dfrac{dz_3}{dt} = \dfrac{20}{2900}z_1(t) + \dfrac{18}{850}z_2(t) - \dfrac{38}{1180}z_3(t) \end{cases} \qquad (6.35)$$

$$\begin{cases} \dfrac{dz_1}{dt} = \dfrac{38}{1180}z_3(t) + 100t - \dfrac{38}{2900}z_1(t) \\[2mm] \dfrac{dz_2}{dt} = \dfrac{8}{2900}z_1(t) - \dfrac{18}{850}z_2(t) \\[2mm] \dfrac{dz_3}{dt} = \dfrac{20}{2900}z_1(t) + \dfrac{18}{850}z_2(t) - \dfrac{38}{1180}z_3(t) \end{cases} \qquad (6.36)$$

$$\begin{cases} \dfrac{dz_1}{dt} = \dfrac{38}{1180}z_3(t) + \{1 + \sin(t)\} - \dfrac{38}{2900}z_1(t) \\[2mm] \dfrac{dz_2}{dt} = \dfrac{8}{2900}z_1(t) - \dfrac{18}{850}z_2(t) \\[2mm] \dfrac{dz_3}{dt} = \dfrac{20}{2900}z_1(t) + \dfrac{18}{850}z_2(t) - \dfrac{38}{1180}z_3(t) \end{cases} \qquad (6.37)$$

with initial conditions $z_1(0) = z_2(0) = z_3(0) = 0$. Z is the amount of the pollutant.

This idea was meant to understand the dynamics of pollutants in lake. In furtherance to understanding the concept of pollutant dynamics in lakes, Shiri and Baleanu (2021) worked on a model that estimates the amount of pollution in lakes connected with some rivers. The assumptions were that the density of pollution in a lake act like a memory. A system of fractional differential equations was derived which was solved using high-order numerical method which was compared with an explicit method based on the regularity of the solution.

Generally, there is not a specific way for forming this kind of model. However, there is much attention on a search for better and more efficient solution methods for determining solution, to physical models. Hence, whatever solution that interest the researcher, i.e., approximate, exact, analytical, or numerical, the emphasis is on the formulation of the model. The system of lake pollution models has been solved first time by Biazar et al. (2006) using a semianalytic technique called the Adomian decomposition method (ADM). Ganji et al. (2018) solved the lake pollution system via homotopy-perturbation method (HPM). El-Dessoky and Altaf (2020) solved dynamics of system of polluted lakes through the numerical procedure called fractional derivative approach. Generally, lake pollution model is formulated when the differential equation is primarily obeys the mass balance of the pollutant, so that

The change in amount of pollutant = Amount entering − Amount leaving.

Other formulations and assumptions are at the scientist discretion. For example, Ghosh et al. (2021) used three-dimensional system of differential equations with three instances of input (i.e., impulse input, step input, and sinusoidal input) to model the pollutants in the lake. A new iterative method was applied to the lake pollution model and compared with the fourth-order Runge—Kutta method (RK4). The knowledge of the numerical analysis discussed in this chapter is of huge importance to a nonmathematical professional in the environmental field because it is widely used and can be used to verify scenarios in any type of field measurements.

Another area of consideration in environmental pollution is air quality. Like the lake pollution, it is an open system and the accuracy of measurement is always compromised. The physics and chemistry of vertical transport processes (turbulent diffusion, advection, deposition) are huge consideration when formulating the differential equations for any part or region of concentration. While academics will concentrate on formulation of numerical techniques, a nonmathematical environmentalist needs the basic understanding on how to solve models either by integrating implicit and explicit method, or applying them differently. Hence, it is important to know the rudiment and by extension the computational requirements to solve problems numerically without stress. In this section, we shall illustrate a scenario of air pollution to show how it is practically investigated.

Air pollution in most parts of the globe has been proven severally to be very high. The implication on the ecosystem is tremendous, as it has led to the death or severe conditions of microorganism. In this paper, the effect of air pollution on buildings was investigated. Air pollution is said to be introduction harmful or poisonous substances into the air which can cause the deterioration of biological life. The pollution introduced into the atmosphere causes contamination of the air we breathe which causes an alteration in the quality of the air which is available. It's mainly accrued to human actions but sometimes result from the occurrence of natural phenomena such as volcanic eruptions, dust storms, and wild fires also add to the depletion of air quality. Some of the major air pollutants include SO_2 which are from fossil fuel combustion for power generation, petroleum refining (Emetere, 2017). The pollution can be gaseous, liquid, or solid and can be classified chemically such as oxide, hydrocarbon, acid, or other kinds of pollutants. The classified as biological entity such as bioaerosols. Most pollutants go into the air from natural sources.

The pollutants are described as either primary pollutant or secondary pollutant. Primary pollutants are directly put into the air. Secondary pollutants are made as a result of chemical reactions when pollutants mix with other primary pollution or natural substances (Envira, 2019). Pollutants are mainly from burning of fossil fuels, emissions of large amounts of carbon monoxides from industries and factories, agricultural activities that introduces harmful chemicals, waste production, mining operations, and indoor pollution. Humans are not the only sole cause of anthropogenic pollution which most of the time occurs naturally because of volcanoes, fires in forests and prairies, dust storms, aerosols over oceans, etc. The effects of air pollution have are described by the extent of the damage caused by the pollutants. Some effects include plant evolution via pollutant deposition on leaves. This activity prevents the occurrence of photosynthesis in many cases that has a consequence on the air we breathe. Also, air pollutant emission into the atmosphere aids the formation of acid rain. Also, it is reported that emission of greenhouse gases causes global warming and climate change. The quantity of CO_2 in the air is a major factor of the greenhouse effect that ought to be beneficial for plants. The continual exposure of human to pollutants is the major cause for the deterioration of human health. Air pollution is a significant risk factor for human health conditions that causes allergies, respiratory and cardiovascular diseases as well as lung damage. Also, the wildlife suffers from pollution that most time deposit on waterways. The presence of toxic chemicals can cause the death of animals that can lead to extinction of their kind. In this study, the focus is to see the reason why significant numbers of building in Basse-Gambia are corroding. The lowest rates of the effects of these pollutants are in cold and dry climates, the intermediate rates are situated in tropics and marine environments while the highest rates of the pollutants are recorded in polluted industrial locations.

The air pollution burden in West Africa is significantly high because of its proximity to Sahara Desert, population, gas flaring, unmonitored industrial emission, bush burning, etc. The research site (Basse) is a town in the Gambia, lying on the south bank of the River Gambia. The beehive of activities encourages significant anthropogenic emission. The air pollution over Basse was investigated in this research using aerosol optical depth data from satellite measurement. The focus of the research is to establish the pollution trend in the area to guide authorities on how to mitigate further emission. Aerosol optical depth (AOD) is a dimensionless number that characterizes the total absorption and scattering effect of particles in the direct or scattered sunlight. Aerosol particles that have been dispersed into the atmosphere are mixed in this layer of the atmosphere. Detection of AOD dataset from satellite remote sensing is agreed to be reliable because it has well-developed methodology, many applications, and extensive literature (Hand et al., 2004).

In this case, the research location is Basse, Gambia (red ring in Fig. 6.1). The geographical coordinates are in latitude and longitude of 13.3094°N and 14.2192°W, respectively. The aerosol optical depth (AOD) dataset was obtained from the Multi-angle Imaging SpectroRadiometer (MISR). The AOD dataset from 2000 to 2013 was used to investigate the air pollution in Basse, Gambia. MISR views the sunlit earth simultaneously at nine widely spaced angles to determine regional and global impacts of different types of atmospheric particles and clouds on climate. MISR operates at Level 2 to retrieve aerosol column amount, aerosol particle properties, and ancillary information based on Level 1B2 geolocated radiances observed by MISR from the National Aeronautics and Space Administration (NASA) Terra Earth Observing System (EOS) satellite. Its dataset is widely used because it is has high reliability.

The AOD dataset is obtained using the Lambert–Beer–Bouguer law:

$$I(\lambda) = I_o(\lambda)\exp(-\tau_a(\lambda)m_t) \tag{6.38}$$

where $I(\lambda)$ is the spectral irradiance at wavelength λ and at the ground level, I, (τ) is the spectral irradiance of the solar beam outside the earth atmosphere, I_o, (λ) is the total extinction (scattering and absorption) optical depth of the atmosphere in the vertical path, and m, is the relative air mass. According to Angstrom's turbidity equation, the aerosol optical depth is the given as (Emetere, 2017):

$$\tau_a(\lambda) = \beta\left(\frac{\lambda}{a_0}\right)^{-\alpha} \tag{6.39}$$

with $a = 1$ pm is simply a scale factor introduced for dimensional consistency. The coefficient β is equal to $\tau_a(\lambda)$ at a wavelength of 1 m and depends on the concentration of particles. Typical values vary from 0 to 0.5; the higher β is, the higher the amount of aerosol present in the atmosphere.

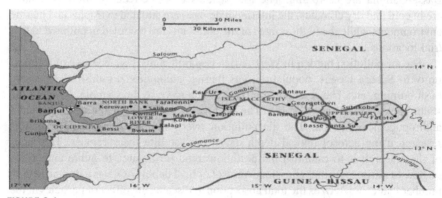

FIGURE 6.1

Location of Basse within Gambia.

The aerosol optical depth (AOD) via the spectral dependence of extinction by particles is given by a power law relationship (Schuster et al., 2006):

$$\tau(\lambda) = \tau_1 \lambda^{-\alpha} \tag{6.40}$$

where $t(l)$ is the aerosol optical depth (AOD) at the wavelength λ, τ_1 is the approximated AOT at a wavelength of 1 mm (sometimes called the turbidity coefficient, as per, and α has come to be widely known as the Angstrom exponent.

Like the lake pollution model, there are aerosol models that could also explain its dynamics in the atmosphere. Christian et al. (1986) gave detailed models for solving aerosols optical depth or distribution problem. For example, the evolution of aerosol size distribution by coagulation is presented as:

$$\frac{\partial n(v,t)}{\partial t} = \frac{1}{2} \int_{v^*}^{v} \beta\left(v - \check{v}, \check{v}\right) n\left(\check{v}, t\right) \times n\left(v - \check{v}, t\right) d\check{v} - \int_{0}^{\infty} \beta\left(v, \check{v}\right) n(v,t) n\left(\check{v}, t\right) d\check{v} \tag{6.41}$$

The initial condition is given as $n(v,0) = n_0(v)$.

where β is the coagulation coefficient, v is the minimum aerosol volume, n is the density function of the aerosol number size distribution, and ndv is the number of aerosol per unit volume of air.

Also, the aerosol size distribution due to vapor condensation growth is presented as:

$$\frac{dv}{dt} = \frac{4\pi r D}{1 + \left(\dfrac{1.333 Kn + 0.71}{1 + Kn}\right) Kn} \times v_m \left[P - P_s \exp\left(\frac{2\sigma v_m}{rKT}\right)\right] \tag{6.42}$$

where r is aerosol radius, D is diffusion coefficient, Kn is Knudsen number, v_m is the molecular volume of the condensing vapor, P is the ambient vapor pressure of the vapor, and P_s is the equilibrium vapor pressure of the vapor, σ is the surface tension of the condensed species, k is the Boltzmann constant, and T is the absolute temperature.

From the satellite measurement, the averages for each of the months between 2000 and 2013 were plotted as presented in Fig. 6.2. It is observed that the AOD have a Boltzmann curve showing that the AOD is lower in dry season than in wet season. This result opposes the general idea that aerosol optical depth is higher during dry season than in wet season (Ayanlade et al., 2019). However this scenario (i.e., higher AOD at wet season than dry season) was found to exist at Kanpur, India, as shown in Fig. 6.3 (Sivaprasad and Babu, 2014). This result was possible because fine mode aerosols are formed by photochemical reactions and biogenic reactions and are predominant during low wind speeds, which is a favorable condition for the formation of these particles. Judging by the Boltzmann curve, unlike Kanpur-India, the AOD in Basse is maximum at April–June (Fig. 6.2) while Kanpur-India at June and July (Fig. 6.3). The third-order polyfit shows that the R-square is 0.89. The coefficients of the polyfit is > 1. The higher AOD for the months are April, June and August. Unlike the atmospheric condition in Pankur-India, coarse mode aerosols are formed by photochemical reactions and biogenic reactions and are predominant during low wind speeds (Fig. 6.4). The agreement of this postulation is supported by the R-square value, i.e., 0.89.

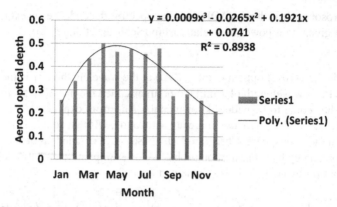

FIGURE 6.2

AOD for Basse-Gambia.

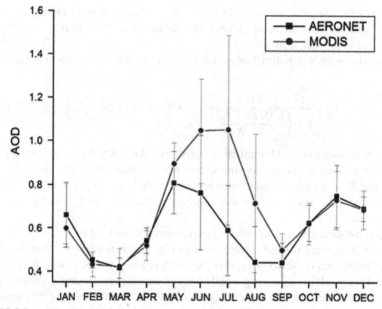

FIGURE 6.3

AOD for Kanpur-India.

The Angstrom exponent is inversely related to the average size of the particles in the aerosol (Fig. 6.5). The R-square of the polyfit is given as 0.92. The Angstrom turbidity parameters can be found using the coefficient that is <1. However, the variation of Angstrom exponent from negative to positive value is indication that the effect of the climate change on the wind system in Basse is evident. Within the period of high AOD, rainfalls in such period are more likely to be acid rain. The impact that

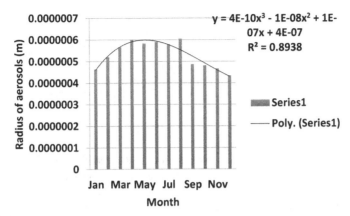

FIGURE 6.4

Radius of aerosols for Basse-Gambia.

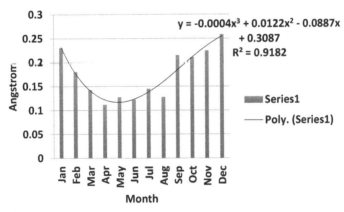

FIGURE 6.5

Angstrom for Basse-Gambia.

the acid deposits have on modern building materials is enormous. Most historic structures are affected by acid rains that at those time studies were carried out. The effects caused by the SO_2 and some other chemicals have major impacts on the buildings that may cause discoloration, material loss, and structural failure. The effects of some of the particulates such as soot, dust, and fumes add to the degradation of fabrics, buildings, and painted surfaces when driven by winds at high speeds and are deposited upon the surface of objects. Due to their corrosiveness added with that of SO_2 quicken the decomposition of the material.

The effects air pollution on building materials have increased over the years and this has caused high cost of building maintenance. The results show that both dry and wet deposition of aerosols, e.g., sulfur dioxide and gaseous pollutants that has been nucleated by water molecule in the atmosphere. Hence, the rate of the corrosion effect is controlled by some strategic factors such as: moisture of the material, type of pollutant deposited on the material and its composition, and temperature. It is recommended that the ways to reduce these pollutants include; transition from fossil fuel to an alternative source of energy, production of clean energy and its conservation is crucial, eco-friendly transportation, and green building. It was observed that coarse mode aerosols are formed by photochemical reactions and biogenic reactions and are predominant during low wind speeds. This revelation proves that gaseous pollutants around the research site are nucleated by water molecule before deposition on the wall or roof the building.

To round-up this section, it is salient to note that research depends on the exposure of the researcher. For example, the above research discussion and analysis can be more matured if the researcher could show the relation between the numerical study and the satellite measurement as shown in the Figure below. Hence, there is the need for a nonmathematician environmentalist to take special interest in the computational analysis of the next subsection.

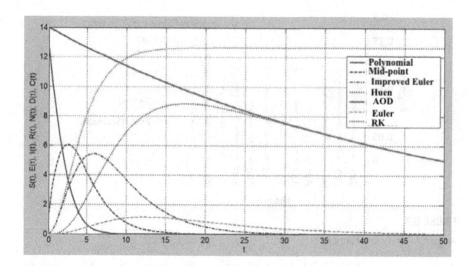

1.7 Computational processing of numerical methods for solving differential equation

1.7.1 The Euler's method

In this section, we shall adapt two techniques to computationally solve the Euler's method using the MATLAB software. We may decide to write an indigenous function as presented below.

Code 6.1: Euler's Function.

```
1.  function E=euler_selfee(f,x,y,z,N)
2.  h=(y-x)/N;
3.  Y=zeros(1,N+1);
4.  T=x:h:y;
5.  Y(1)=z;
6.  for j=1:N
7.  Y(j+1)=Y(j)+h*f(T(j));
8.  end
9.  E=[T' Y'];
10. End
```

where f is the function entered as function handle; x and y are the left and right end-points; z is the initial condition $E(a)$; N is the number of steps; $E = [T'\ Y']$, where T is the vector of abscissas and Y is the vector of ordinates.

Code 6.2: Euler's code.

```
    %Define the differential equation
1.  f=@(x) (y*log(y))/(2*x);
    %List the component of the initial condition
2.  x=2;
3.  y=2.7182818;
    %In the finction (euler_selfee, find the size step
4.  b=10;
5.  N=200;
    % find the euler to find 'x' and 'y'
6.  y1=euler_selfee(f,x,b,y,N);
    %plot the component of the euler
7.  plot(y1(:,1),y1(:,2),'b','linewidth',3);
8.  xlabel('X');
9.  ylabel('Y');
    % Increase the font in both axes
10. set(gca,'FontSize',18);
```

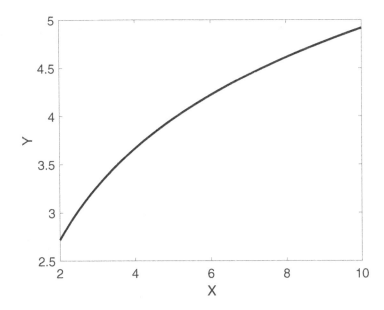

To avoid the stress of writing your own function, you can adapt the function of Euler written by Lakhdar (2022) titled "euler.m" which was modified to enable plot as presented in Code6.3. Code 6.4 is the plot code for the Euler.

Code 6.3: Euler's function2.

```
1.  function [y,t]=euler(f,y0,a,b,h)
2.  y(1)=y0;
3.  t(1)=a;
4.  n=(b-a)/h;
5.  for k=1:h
6.  y(k+1)=y(k)+n*f(t(k));
7.  t(k+1)=t(k)+n;
8.  end
```

Code 6.4: Euler's function2.

```
    %Define the differential equation
1.  f=@(x) (y*log(y))/(2*x);
    %List the component of the initial condition
2.  x=2;
3.  y=2.7182818;
    %In the finction (euler_selfee, find the size step
4.  b=10;
5.  N=200;
    % find the euler to find 'x' and 'y'
6.  [y1,t1]=euler(f,y,x,b,N);
    %plot the component of the euler
7.  plot(t1,y1,'b','linewidth',3);
8.  xlabel('X');
9.  ylabel('Y');
    % Increase the font in both axes
10. set(gca,'FontSize',18);
```

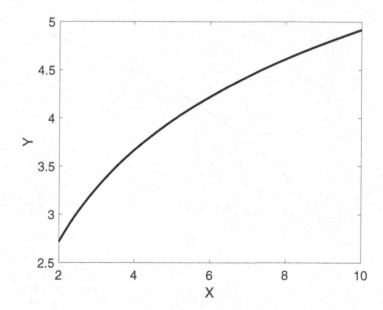

Recall that we had earlier mentioned that the Euler's method will be used to solve higher order differential equation, hence, the example below.

Solve $y''' = 2y'' + xy' - 6y$,

The initial conditions are $y(0) = 1$, $y'(0) = 0$, $y''(0) = 1$, interval is $[0, 5]$, step size $= 0.001$.

Code 6.5: Euler's code to solve higher order differential equation.

```
11. clear
12. close all
13. clc
%initial value
14. t0 = 0;
%initial condition(given)
15. y0 = 1;
%end of time step
16. tEnd = 10;
%step size
17. h = 0.01;
%number of interval
18. N = (tEnd-t0)/h;
19. T = t0:h:tEnd;
%define the derivative
20. y(1) = y0;
21. y(2) = y(1)';
22. y(3) = y(2)';
23. y(4) = y(3)';
%Define the function
24. y(4)= 2*y(3) + T(:,1)*y(2)-6*y(1);
% a series of 3-element column vectors
25. y = zeros(3, N+1);
% initial value is a 3-element column vector of [ y ; y' ; y" ]
26. y(:,1) = [1;0;1];
27. for i = 1:N
%function, which needs to calculate a 3-element derivative column vector
28. fi = [y(1)'; y(2)'; y(3)'];
% T(i+1) = T(i) + h;
29. y(:,i+1) = y(:,i) + h * fi*T(i);
30. end
%ploting the graph
31. plot(T,y(3,:),'b','LineWidth',3);
32. xlabel('X');
33. ylabel('Y');
% Increase the font in both axes
34. set(gca,'FontSize',18);
```

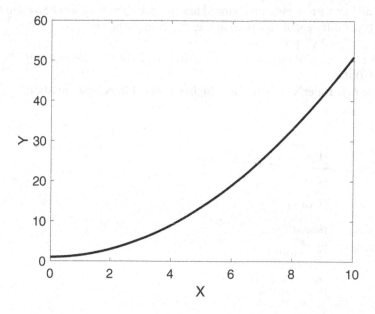

Next is the coding for the predictor corrector of same problem displayed in the Euler. The code was modified based on existing code by ThemeXpose (2022) as presented in code6.6.

Code 6.5: Predictor Corrector Method.

```
%Write differential equation to be solved
  1.  f = @(t,y) (y*log(y))/(2*t);
%list the parameters i.e., a is the left end point, b is the right end point, n is the no. of
subintervals, alpha is the initial condition
  2.  a = 2;
  3.  b = 10;
  4.  n = 200;
  5.  alpha =2.7182818;
%calculate the step size
  6.  h = (b-a)/n;
  7.  t(1) = a;
  8.  w(1) = alpha;
  9.  fprintf('   t       w\n');
  10. fprintf('%5.4f %11.8f\n', t(1), w(1));
% predictor calculation
  11. for i = 1:3
  12. t(i+1) = t(i)+h;
  13. k1 = h*f(t(i), w(i));
  14. k2 = h*f(t(i)+0.5*h, w(i)+0.5*k1);
  15. k3 = h*f(t(i)+0.5*h, w(i)+0.5*k2);
  16. k4 = h*f(t(i+1), w(i)+k3);
  17. w(i+1) = w(i)+(k1+2.0*(k2+k3)+k4)/6.0;
  18. fprintf('%5.4f %11.8f\n', t(i+1), w(i+1));
  19 end
% corrector calculation
  20. for i = 4:n
  21. t0 = a+i*h;
  22. part1 = 55.0*f(t(4),w(4))-59.0*f(t(3),w(3))+37.0*f(t(2),w(2));
  23. part2 = -9.0*f(t(1),w(1));
  24. w0 = w(4)+h*(part1+part2)/24.0;
  25. part1 = 9.0*f(t0,w0)+19.0*f(t(4),w(4))-5.0*f(t(3),w(3))+f(t(2),w(2));
  26. w0 = w(4)+h*(part1)/24.0;
  27. fprintf('%5.4f %11.8f\n', t0, w0);
  28. for j = 1:3
  29. t(j) = t(j+1);
  30. w(j) = w(j+1);
  31. end
  32. t(4) = t0;
  33. w(4) = w0;
  34. end
%ploting the graph
  35. plot(t,w,'b','LineWidth',3);
  36. xlabel('X');
  37. ylabel('Y');
% Increase the font in both axes
  38. set(gca,'FontSize',18);
```

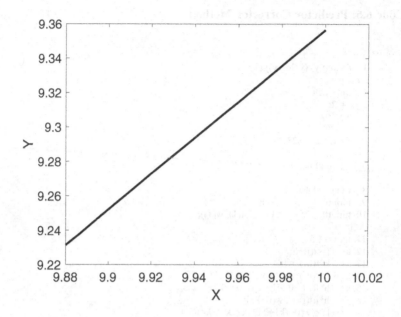

Comparing the Euler and the predictor-corrector graphs, it is clear that their convergence differs. Hence, the choice of any depends on what the user expects to see. The application and usage of the Runge–Kutta has been fully illustrated by Tillmann (2022).

1.8 Computational application of derivatives to environmental data

Oza (2018) proposed useful derivatives of data extraction and big data concepts. This research is an insight to the enormous opportunities that environmental researcher have to transform their observations to worthwhile information. Simply put, a time shall come where formulation of mathematical model may no longer be in vogue in environmental science. With the concept of big data, environmental researcher have lots of opportunity to process "Big Data" at same time with the data extraction from numerous different sources, including satellite dataset. In this section, we consider the different ways the derivative of dataset can be obtained or derived computationally and executed numerically.

1.9 Case 1: derivative of experimental data

The dataset used for this section is presented below. The code for the first dataset is presented in Code6.6. The second code is for the combination of the two dataset.

Code6.6.

```
%define the names of all the excel file
1.  data=xlsread('exp.xls');
2.  x = data(:,1);
3.  y = data(:,2);
    %Find the derivative of the data
4.  dydx = diff([eps; y(:)])./diff([eps; x(:)]);
    %Plot the dataset
5.  figure
6.  plot(x,dydx,x,y,'linewidth',3);
7.  ylabel('y');
8.  xlabel('x');
9.  legend('Derivative', 'Experimental');
10. set(gca,'FontSize',18);
```

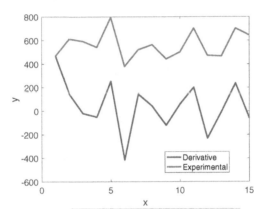

A	B	C
1	467.79	339.25
2	612.47	386.8
3	592.9	323.33
4	542.32	535.35
5	795.11	807.6
6	379.32	175.07
7	521.41	338.95
8	563.21	391.58
9	442.82	137.01
10	502.23	309.87
11	703.11	620.69
12	472.2	186.19
13	466.6	308.43
14	702.72	625.37
15	643.45	401.13

The second example was to find the first derivative of two sets of data. This section is important if the research wished to have a comparative analysis of the data set.
Code6.7.

```
%define the names of all the excel
data=xlsread('exp.xls');
x = data(:,1);
y = data(:,2);
y1 = data(:,3);
%Find the derivative of the data
dydx = diff([eps; y(:)])./diff([eps; x(:)]);
dy1dx = diff([eps; y1(:)])./diff([eps; x(:)]);
%Plot the dataset
figure
plot(x,dydx,'-b',x,dy1dx,'r','linewidth',3);
ylabel('y');
xlabel('x');
legend('Derivative-data1', 'Derivative-data2');
set(gca,'FontSize',18)
```

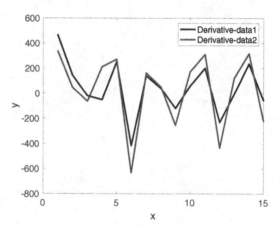

The derivation of the polynomic function of the derivative is presented below and the coefficients displayed in a table. The coefficient table is important to synthesize the boundary conditions in live scenarios as proven by Samalerk and Pochai (2018).

Code6.8.

```
%define the names of all the excel files
  1. data=xlsread('exp.xls');
  2. x = data(:,1);
  3. y = data(:,2);
  4. y1 = data(:,3);
%Find the derivative of the data
  5. dydx = diff([eps; y(:)])./diff([eps; x(:)]);
  6. dy1dx = diff([eps; y1(:)])./diff([eps; x(:)]);
%Find the polyfit of function for derivative
  7. p = polyfit(x,dydx,14);
  8. p2 = polyfit(x,dy1dx,14);
  9. y2 = polyval(p,x);
  10. y3 = polyval(p2,x);
%Find the polyfit of function for derivative
  11. p3 = polyfit(x,y,14);
  12. p4 = polyfit(x,y1,14);
%Plot the dataset
  13. figure
  14. plot(x,dydx,'bo',x,y2,'r-',x,dy1dx,'go',x,y3,'m-','linewidth',3)
  15. ylabel('y');
  16. xlabel('x');
  17. legend('Derivative-data1','Derivative-polyfit1','Derivative-data1', 'Derivative-
      polyfit2');
  18. set(gca,'FontSize',18)
```

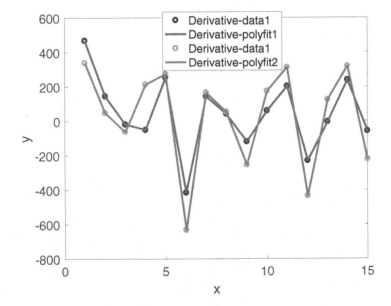

Coefficient table.

Coeff_Der_1	Coeff_data1	Coeff_Der_2	Coeff_data2
1.36E-05	4.67E-06	1.30E-05	3.74E-06
−0.0015515	−0.0005344	−0.0014933	−0.0004325
0.08041795	0.02777024	0.07795435	0.02277933
−2.5031389	−0.867,564	−2.4442554	−0.7221609
52.1613052	18.1653158	51.3098148	15.3588098
−767.86459	−269.00208	−760.81179	−231.16355
8210.90335	2896.84789	8192.20518	2530.6511
−64,567.96	−22,964.883	−64,840.029	−20,389.78
373,691.636	134,106.767	377,466.212	120,929.826
−1,576,352	−571,151.14	−1,600,334.6	−522,465.29
4,742,837.18	1,735,486.47	4,834,970.94	1,607,897.33
−9,798,118.6	−3,620,365	−10,020,352	−3,390,712.1
13,017,290.2	4,853,795.92	13,342,999.1	4,585,963.17
−9,833,407.5	−3,695,638.6	−10,094,736	−3,515,485.1
3,131,602.06	1,184,553.07	3,217,685.02	1,132,287.05

Creating a function from a given dataset is demonstrated below. This means that though the mathematical display cannot be displayed, the function enables numerical analysis of all sorts, i.e., numerical differentiation, integration, and interpolation. The second-order derivative was obtained in this example using the computationally crafted function.

Code6.9:

```
%define the names of all the excel files
data=xlsread('exp.xls');
x = data(:,1);
y = data(:,2);
%Create the function of the graph
F = griddedInterpolant(x,y);
fun = @(x) F(x);
Y = diff([eps; fun(x)]);   % first derivative
Z = diff([eps; Y(:)]);     % second derivative
%Plot the graph
plot(x,Z,'r-',x,Y,'m-',x,fun(x),'b-','linewidth',3)
ylabel('y');
xlabel('x');
legend('Second-Derivative-data1','First-Derivative-data1','Function-data1');
set(gca,'FontSize',18)
```

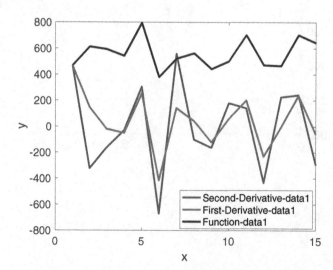

References

Aguirre, J., Tully, D., 2019. Lake Pollution Model. https://mse.redwoods.edu/darnold/math55/DEProj/Sp99/DarJoel/lakepollution.pdf. (Accessed 5 February 2022).

Ayanlade, A., Godwin, A., Jegede, M.O., 2019. Spatial and seasonal variations in atmospheric aerosols over Nigeria: assessment of influence of intertropical discontinuity movement. Int. J. Ocean Clim. Syst. 9, 1–13.

Biazar, J., Farrokhi, L., Islam, M.R., 2006. Modeling the pollution of a system of lakes. Appl. Math. Comput. 178 (2), 423–430.

Christian, S., Hudischewskyj, A.B., Seinfeld, J.H., Whitby, K.T., Whitby, E.R., James, R.B., Barnes, H.M., 1986. Simulation of aerosol dynamics: a comparative review of mathematical models. Aerosol. Sci. Technol. 5 (2), 205–222. https://doi.org/10.1080/02786828608959088.

El-Dessoky, A.M.M., Altaf, K.M., 2020. Modeling and analysis of the polluted lakes system with various fractional approaches. Chaos Solit. Fractals 134, 109720.

Emetere, M.E., 2017. Investigations on aerosols transport over micro- and macro-scale settings of West Africa. Environ. Eng. Res. 22 (1), 75–86.

Envira, 2019. Primary and Secondary Pollutants: These Are the Most Dangerous. https://enviraiot.com/primary-and-secondary-pollutants-most-dangerous/.

Ganji, D.D., Bandpy, M.G., Mostofi, M., 2018. Simulation of lakes pollution distribution by homotopy perturbation method. Int. J. Eng. Appl. (IREA) 6 (2), 52–56.

Ghosh, I., Chowdhury, M.S.H., Aznam, S.M., Rashid, M.M., 2021. Measuring the pollutants in a system of three interconnecting lakes by the semianalytical method. J. Appl. Math. 2021, 16. https://doi.org/10.1155/2021/6664307. Article ID 6664307.

Hand, J.L., Kreidenweis, S.M., Slusser, J., Scott, G., 2004. Comparison of aerosol optical properties derived from sun photometry to estimates inferred from surface measurements in Big Bend National Park, Texas. Atmos. Environ. 38, 6813–6821.

Lakhdar, A., 2022. Euler Function. https://www.mathworks.com/matlabcentral/fileexchange/105070-euler?s_tid=srchtitle.

Mungkasi, S., 2021. Variational iteration and successive approximation methods for a SIR epidemic model with constant vaccination strategy. Appl. Math. Model. 90, 1–10.

Oza, H., 2018. Some Useful Derivatives of Data Extraction and Big Data Concepts. https://www.dataversity.net/useful-derivatives-data-extraction-big-data-concepts/.

Prakashaa, D.G., Veereshab, P., 2020. Analysis of Lakes pollution model with Mittag-Leffler kernel. J. Ocean Eng. Sci. 5 (4), 310–322.

Samalerk, P., Pochai, N., 2018. Numerical simulation of a one-dimensional water-quality model in a stream using a saulyev technique with quadratic interpolated initial-boundary conditions. Abstr. Appl. Anal. 1926519. https://doi.org/10.1155/2018/1926519.

Schuster, G.L., Dubovik, O., Holben, B.N., 2006. Angstrom exponent and bimodal aerosol size distributions. J. Geophys. Res. 111, D07207.

Shiri, B., Baleanu, D., 2021. A general fractional pollution model for lakes. Commun. Appl. Math. Comput. https://doi.org/10.1007/s42967-021-00135-4.

Sivaprasad, P., Babu, C.A., 2014. Seasonal variation and classification of aerosols over an inland station in India. Meteorol. Appl. 21, 241–248.

ThemeXpose, 2022. Predictor Corrector Method Using MATLAB. https://www.matlabcoding.com/2019/02/predictor-corrector-method-using-matlab.html.

Tillmann, S., 2022. Runge-Kutta fixed step solvers. In: MATLAB Central File Exchange. Retrieved. https://www.mathworks.com/matlabcentral/fileexchange/69951-runge-kutta-fixed-step-solvers. (Accessed 6 February 2022).

Further reading

Peiravi, S., 2019. Numerical study of fins arrangement and nanofluids effects on three-dimensional natural convection in the cubical enclosure. Transp. Phenom. Nano Micro Scales 7 (2), 97–112.

Torsten, 2018. 2nd Order ODE Using Euler Method. https://www.mathworks.com/matlabcentral/answers/431826-2nd-order-ode-using-euler-method.

Numerical integration application to environmental data

1. Introduction

In this chapter, the types of numerical integration techniques and its application to environmental dataset was discussed. Many integral equations are difficult to solve or complex to solve analytically because of the complexity of its function. The shortest route to solving such challenges is to adopt various techniques of numerical integration to approximate their values. In this chapter, the Midpoint, Trapezoidal, and Simpson's rule will be considered as a numerical integration technique. It is important to note that these numerical integration techniques has its absolute and relative errors which varies depending on the integral function. Absolute error is the difference between the actual value and the calculated value while the relative error is the ratio of the absolute error and the experimental value.

1.1 Midpoint

The midpoint rule is identified as a popular numerical integral technique for estimating a definite integral. It works on the rudiment of the Riemann sum. The Riemann sum entails the estimation of the subintervals of equal width to obtain its midpoints. The summation of the approximate solutions of individual midpoints gives the solution of the definite integral.

In principle, the formulation of the Midpoint rule assumes a function denoted as $f(x)$ which is continuous along interval $[b, c]$. Hence, if the subintervals are denoted as n, then the length of each subintervals is calculated as

$$\Delta x = \frac{c - b}{n}$$

The midpoint of the jth subinterval is written as

$$M_n = \sum_{j=1}^{n} f(m_j) \Delta x$$

Numerical Methods in Environmental Data Analysis. https://doi.org/10.1016/B978-0-12-818971-9.00008-9

This formula simply denotes that if $f(x) > 0$, then the approximate solution of the integral

$$\lim_{n \to \infty} M_n = \int_b^c f(x)dx$$

is the corresponding sum of the areas of rectangles approximating the area between the graph of $f(x)$ and the x-axis over $[b,c]$ at a subinterval of 4 which is represented in the figure below.

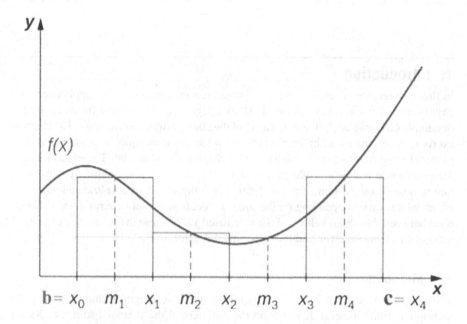

The above graph shows an integral with four subintervals. The rectangles of the chart corresponds to M4 for a nonnegative function over a closed interval $[b,c]$. The modalities of the midpoint rule is demonstrated in the example below.

Use the midpoint to estimate $\int_1^3 3x^2 dx$ using 10 intervals.

First the divisions of the intervals are calculated using the subintervals

$$\Delta x = \frac{3-1}{10} = \frac{1}{5}$$

The divisions of the subinterval are written as $\left[1, 1\frac{1}{5}\right]$, $\left[1\frac{1}{5}, 1\frac{2}{5}\right]$, $\left[1\frac{2}{5}, 1\frac{3}{5}\right]$, $\left[1\frac{3}{5}, 1\frac{4}{5}\right]$, $\left[1\frac{4}{5}, 2\right]$, $\left[2, 2\frac{1}{5}\right]$, $\left[2\frac{1}{5}, 2\frac{2}{5}\right]$, $\left[2\frac{2}{5}, 2\frac{3}{5}\right]$, $\left[2\frac{3}{5}, 2\frac{4}{5}\right]$, and $\left[2\frac{4}{5}, 3\right]$.

The second step is finding the midpoints of this intervals

$$\left\{ \frac{11}{10}, \frac{13}{10}, \frac{15}{10}, \frac{17}{10}, \frac{19}{10}, \frac{21}{10}, \frac{23}{10}, \frac{25}{10}, \frac{27}{10}, \frac{29}{10} \right\}$$

The third step is to calculate the approximate for each midpoint

$$M_{10} = \frac{1}{5} \cdot f\left(\frac{11}{10}\right) + \frac{1}{5} \cdot f\left(\frac{13}{10}\right) + \frac{1}{5} \cdot f\left(\frac{15}{10}\right) + \frac{1}{5} \cdot f\left(\frac{17}{10}\right) + \frac{1}{5} \cdot f\left(\frac{19}{10}\right) + \frac{1}{5} \cdot f\left(\frac{21}{10}\right)$$

$$+ \frac{1}{5} \cdot f\left(\frac{23}{10}\right) + \frac{1}{5} \cdot f\left(\frac{25}{10}\right) + \frac{1}{5} \cdot f\left(\frac{27}{10}\right) + \frac{1}{5} \cdot f\left(\frac{29}{10}\right)$$

$$= \frac{363}{500} + \frac{507}{500} + \frac{675}{500} + \frac{867}{500} + \frac{1083}{500} + \frac{1323}{500} + \frac{1587}{500} + \frac{1875}{500} + \frac{2187}{500} + \frac{2523}{500}$$

$$M_{10} = \frac{12990}{500} \approx 25.98$$

When this solution is compared with the actual value of the definite integral, the approximation error in the midpoint rule is large. However, this error depends on the function. The error in the midpoint rule is calculated with respect to absolute and relative error.

From the above example, the absolute error is given as:

$E =$ value of the integral function-value of the midpoint

$$E = |26 - 25.98| \approx 13.01$$

The relative error is calculated as:

$$R = \frac{Absolute\ error\ (E)}{Exact\ value} = \frac{0.02}{26}$$

$$R \approx 0.000769 \approx 0.0769\%$$

The error bound for the midpoint rule is formulated on the assumption that the function $f(x)$ is continuous over the interval $[b,c]$. In that scenario, the maximum value of the second derivative $f'(x)$ over $[b,c]$ is used to estimate the upper bound. Hence, error bound for the midpoint rule is given as:

$$M_n \leq \frac{M(c - b)^3}{24n^2}$$

1.2 Trapezoidal rule

This method is a popular numerical integration technique. It works on the ideology that the approximation of a definite integral solution can be achieved using trapezoids rather than rectangles. This means the area under the curve of the definite integral must satisfy the properties of a trapezoids. A trapezoid is a quadrilateral (a shape with four sides) with exactly one pair of parallel sides (the parallel sides are called bases) whose base bases are parallel by definition and each lower base angle is supplementary to the upper base angle on the same side (see figure below). The area under the curve is calculated using the formula for the area of a trapezium i.e.,

$$Area = \frac{1}{2} \times h(b_2 + b_1)$$

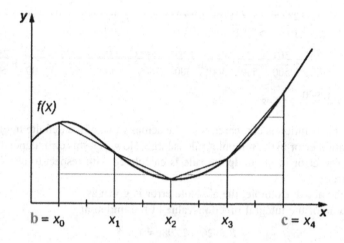

In order to gain insight into the final form of the rule, the length of each subinterval is given by Δx. From the above diagram, the first trapezoid has a height Δx and parallel bases of length (x_0) and (x_1). The height of each subinterval is calculated as:

$$\Delta x = \frac{c - b}{n}$$

The second, third and fourth trapezoid has same height i.e., Δx with bases of $(f(x_2), f(x_1))$, $(f(x_3), f(x_2))$, and $(f(x_4), f(x_3))$ respectively. Hence, the summation of the areas under the curve is given as:

$$\int_a^b f(x)dx \approx \frac{1}{2}\Delta x(f(x_0) + f(x_1)) + \frac{1}{2}\Delta x(f(x_1) + f(x_2)) + \frac{1}{2}\Delta x(f(x_2) + f(x_3)) + \frac{1}{2}\Delta x(f(x_3) + f(x_4))$$

When $\frac{1}{2}\Delta x$ is factorized in the above equation, the trapezoid rule for numerical integration becomes:

$$\int_a^b f(x)dx \approx \frac{1}{2}\Delta x(f(x_0) + 2f(x_1) + 2f(x_2) + 2f(x_3) + f(x_4))$$

The modalities of the midpoint rule are demonstrated in the example below.

Use the midpoint to estimate $\int_1^3 3x^2 dx$ using 10 intervals.

First the divisions of the intervals are calculated using the subintervals

$$\Delta x = \frac{3-1}{10} = \frac{1}{5}$$

The endpoints of the subintervals consist of elements of the set

$$\left\{ 1, 1\frac{1}{5}, 1\frac{2}{5}, 1\frac{3}{5}, 1\frac{4}{5}, 2, 2\frac{1}{5}, 2\frac{2}{5}, 2\frac{3}{5}, 2\frac{4}{5}, 3 \right\}$$

$$T_n \approx \frac{1}{2}\Delta x (f(x_0) + 2f(x_1) + 2f(x_2) + 2f(x_3) + f(x_4))$$

$$= \frac{1}{2} \cdot \frac{1}{5} \left(f(1) + 2f\left(1\frac{1}{5}\right) + 2f\left(1\frac{2}{5}\right) + 2f\left(1\frac{3}{5}\right) + 2f\left(1\frac{4}{5}\right) + 2f(2) + 2f\left(2\frac{1}{5}\right) \right.$$
$$\left. + 2f\left(2\frac{2}{5}\right) + 2f\left(2\frac{3}{5}\right) + 2f\left(2\frac{4}{5}\right) + f(3) \right)$$

$$= \frac{1}{2} \cdot \frac{1}{5} \left(f(1) + 2f\left(\frac{6}{5}\right) + 2f\left(\frac{7}{5}\right) + 2f\left(\frac{8}{5}\right) + 2f\left(\frac{9}{5}\right) + 2f(2) + 2f\left(\frac{11}{5}\right) \right.$$
$$\left. + 2f\left(\frac{12}{5}\right) + 2f\left(\frac{13}{5}\right) + 2f\left(\frac{14}{5}\right) + f(3) \right)$$

$$= \frac{1}{2} \cdot \frac{1}{5} \left(3 + 2 \cdot \left(\frac{108}{25}\right) + 2\left(\frac{147}{25}\right) + 2\left(\frac{192}{25}\right) + 2\left(\frac{243}{25}\right) + 2 \times 12 + 2\left(\frac{363}{25}\right) \right.$$
$$\left. + 2\left(\frac{432}{25}\right) + 2\left(\frac{507}{25}\right) + 2\left(\frac{588}{25}\right) + 27 \right)$$

$$= \frac{1}{10} \left(3 + \left(\frac{216}{25}\right) + \left(\frac{294}{25}\right) + \left(\frac{384}{25}\right) + \left(\frac{486}{25}\right) + 24 + \left(\frac{726}{25}\right) + \left(\frac{864}{25}\right) + \left(\frac{1014}{25}\right) + \left(\frac{1176}{25}\right) + 27 \right)$$

$$= \frac{1}{10} \left(\frac{6510}{25}\right) = 26.04$$

The absolute error is given as:
E = value of the integral function-value of the midpoint

$$E = |26 - 26.04| \approx 0.04$$

The relative error is calculated as:

$$R = \frac{Absolute\ error\ (E)}{Exact\ value} = \frac{0.04}{26}$$

$$R \approx -0.0015 \approx -0.15\%$$

It can be observed that the results of the trapezoid rule is more accurate compared to the midpoint rule. Although most of the times, trapezoidal rule tends to overestimate the value of a definite integral systematically over intervals where the function is concave up and to underestimate the value of a definite integral systematically over intervals where the function is concave down (Openstax, 2022). This statement is both true to both methods as it basically depends on the function of the integral. Though few persons believe that the midpoint rule is more accurate than the trapezoid rule but it is relative to the function and the subintervals in the solution.

The error bound for the trapezoidal rule is formulated on the assumption that the function $f(x)$ is continuous over the interval $[b,c]$. In that scenario, the maximum value of the second derivative $f'(x)$ over $[b,c]$ is used to estimate the upper bound. Hence, error bound for the trapezoidal rule is given as:

$$T_n \leq \frac{M(c-b)^3}{12n^2}$$

1.3 Simpson's rule

There are two main Simpson's rules i.e., Simpson's 1/3 and 3/8 rules. The basic and most popular rule is the Simpson's 1/3 rule. Unlike midpoint and trapezoid rules, Simpson's rules involve several approximations to obtain the solutions of definite integrals. In the midpoint rule, the area under the curve was estimated using piecewise constant functions. The trapezoidal rule on the other hand makes use of piecewise linear functions and the Simpson's rule makes use of piecewise quadratic functions to estimate the area under the curve.

We shall be considering the Simpson's 1/3 which works when 1/3 rule is applied to n equal subdivisions of the integration range $[b, c]$ to get the composite Simpson's rule (See figure below). In this case, points inside the integration range are given alternating weights 4/3 and 2/3.

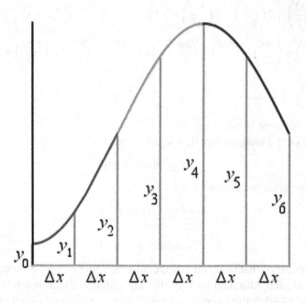

In Simpson's Rule, parabolas are used to approximate each part of the curve. For example, the figure above presents a minimum of three parabolas i.e., along

y_1 and y_2, along y_3 and y_4, along y_5 and y_6. This process is believed to be very efficient since based on a very low relative error when compared with the midpoint and trapezoid rules.

Like the midpoint and trapezoid rules, the formulation of the Simpson's rule also works on the assumption that $f(x)$ is continuous along interval $[b,c]$. Hence, length of each subinterval (n) over $[b,c]$ is given as:

$$\Delta x = \frac{c-b}{n}$$

Let us adopt the a replacement of the function $f(x)$ with $p(x)$ such that

$$S_1 = \int_{x_0}^{x_2} p(x)dx \tag{7.1}$$

where a quadratic function $p(x) = Ax^2 + Bx + C$ is meant to describe the parabola along $(x_0, f(x_0))$, $(x_1, f(x_1))$, and $(x_2, f(x_2))$.where:

$$f(x_0) = p(x_0) = Ax_0^2 + Bx_0 + C \tag{7.2}$$

$$f(x_1) = p(x_1) = Ax_1^2 + Bx_1 + C \tag{7.3}$$

$$f(x_2) = p(x_2) = Ax_2^2 + Bx_2 + C \tag{7.4}$$

The length is $x_2 - x_0 = 2\Delta x$ since $x_1 - x_0 = \Delta x$ and $x_2 - x_1 = \Delta x$. Also, these formulae can be used to find the relationship between x_0, x_1, and x_2

$$x_2 - x_0 = 2(x_1 - x_0)$$

$$x_1 = \frac{x_2 + x_0}{2} \tag{7.5}$$

Insert the quadratic function into Eq. (7.1)

$$= \int_{x_0}^{x_2} \left(Ax^2 + Bx + C\right)dx$$

$$\left(\frac{A}{3}x^3 + \frac{B}{2}x^2 + Cx\right)\Big|_{x_0}^{x_2}$$

$$\frac{A}{3}\left(x_2^3 - x_0^3\right) + \frac{B}{2}\left(x_2^2 - x_0^2\right) + C(x_2 - x_0)$$

$$\frac{A}{3}(x_2 - x_0)\left(x_2^2 + x_2x_0 + x_0^2\right) + \frac{B}{2}(x_2 - x_0)(x_2 + x_0) + C(x_2 - x_0)$$

Factorize $\frac{(x_2 - x_0)}{6}$ from the above equation

$$\frac{(x_2 - x_0)}{6}\left[2A\left(x_2^2 + x_2x_0 + x_0^2\right) + 3B(x_2 + x_0) + 6C\right]$$

Categorize into the ab-initio quadratic function and apply Eqs. (7.4) and (7.2)

$$= \frac{(x_2 - x_0)}{6} \left[(Ax_2^2 + Bx_2 + C) + (Ax_0^2 + Bx_0 + C) + A(x_2^2 + 2x_2x_0 + x_0^2) + 2B(x_2 + x_0) + 4C \right]$$

$$= \frac{\Delta x}{3} \left[(f(x_2) + f(x_0) + A(x_2^2 + 2x_2x_0 + x_0^2) + 2B(x_2 + x_0) + 4C \right]$$

$$= \frac{\Delta x}{3} \left[(f(x_2) + f(x_0) + A(x_2 + x_0)^2 + 2B(x_2 + x_0) + 4C \right]$$

Insert Eq. (7.5)

$$= \frac{\Delta x}{3} \left[(f(x_2) + f(x_0) + A(2x_1)^2 + 2B(2x_1) + 4C \right]$$

$$= \frac{\Delta x}{3} \left[(f(x_2) + f(x_0) + 4Ax_1^2 + 4Bx_1 + 4C \right]$$

Factorize 4 out of the last three terms

$$= \frac{\Delta x}{3} \left[(f(x_2) + f(x_0) + 4(Ax_1^2 + Bx_1 + C) \right]$$

$$= \frac{\Delta x}{3} \left[(f(x_2) + 4f(x_1) + f(x_0) \right]$$

Considering the second parabola,

$$S_2 = \int_{x_2}^{x_4} p(x) dx$$

where a quadratic function $p(x) = Ax^2 + Bx + C$ is meant to describe the parabola along $(x_2, f(x_2))$, $(x_3, f(x_3))$, and $(x_4, f(x_4))$.

$$f(x_2) = p(x_2) = Ax_2^2 + Bx_2 + C$$

$$f(x_3) = p(x_3) = Ax_3^2 + Bx_3 + C$$

$$f(x_4) = p(x_4) = Ax_4^2 + Bx_4 + C$$

Applying the process established within intervals (x_0, x_2) gives

$$= \frac{\Delta x}{3} \left[(f(x_4) + 4f(x_3) + f(x_2) \right]$$

Considering the third parabola,

$$S_3 = \int_{x_4}^{x_6} p(x) dx$$

where a quadratic function $p(x) = Ax^2 + Bx + C$ is meant to describe the parabola along $(x_4, f(x_4))$, $(x_5, f(x_5))$, and $(x_6, f(x_6))$.

$$f(x_4) = p(x_4) = Ax_4^2 + Bx_4 + C$$

$$f(x_5) = p(x_5) = Ax_5^2 + Bx_5 + C$$

$$f(x_6) = p(x_6) = Ax_6^2 + Bx_6 + C$$

Applying the process established within intervals (x_0, x_2) gives

$$= \frac{\Delta x}{3}[(f(x_6) + 4f(x_5) + f(x_4)]$$

Therefore, combining the three parabolas gives

$$= \int_{x_0}^{x_6} p(x)dx = \frac{\Delta x}{3}[f(x_0) + 4f(x_1) + 2f(x_2) + 4f(x_3) + 2f(x_4) + 4f(x_5) + f(x_6)]$$

Hence, the general equation for the Simpson's rule becomes

$$= \int_{x_0}^{x_6} p(x)dx = \frac{\Delta x}{3}[f(x_0) + 4f(x_1) + 2f(x_2) + 4f(x_3) + 2f(x_4) + \ldots\ldots$$
$$+ 2f(x_{n-2}) + 4f(x_{n-1}) + f(x_n)]$$

Like the midpoint and trapezoidal rules, the error bound for the Simpson's rule is formulated on the assumption that the function $f(x)$ is continuous over the interval $[b,c]$. In that scenario, the maximum value of the second derivative $f'(x)$ over $[b,c]$ is used to estimate the upper bound. Hence, error bound for the trapezoidal rule is given as:

$$S_n \leq \frac{M(c-h)^5}{180n^4}$$

Use the Simpson's rule to estimate $\int_1^3 3x^2 dx$ using five intervals.
First the divisions of the intervals are calculated using the subintervals

$$\Delta x = \frac{3-1}{5} = \frac{2}{5}$$

The endpoints of the subintervals consist of elements of the set

$$\left\{ 1, \ 1\frac{2}{5}, \ 1\frac{4}{5}, \ 2\frac{1}{5}, \ 2\frac{3}{5}, 3 \right\}$$

$$= \int_1^3 p(x)dx - \frac{\Delta x}{3}[f(x_0) + 4f(x_1) + 2f(x_2) + 4f(x_3) + 2f(x_4) + 4f(x_5) + f(x_6)]$$

Since $f(x_0) = 0$

$$= \frac{1}{3} \times \frac{2}{5}\left[4f(1) + 2f\left(1\frac{2}{5}\right) + 4f\left(1\frac{4}{5}\right) + 2f\left(2\frac{1}{5}\right) + 4f\left(2\frac{3}{5}\right) + f(3)\right]$$

$$= \frac{1}{3} \times \frac{2}{5}\left[4 \times 3 + 2 \times \frac{147}{25} + 4 \times \frac{243}{25} + 2 \times \frac{363}{25} + 4 \times \frac{507}{25} + 27\right]$$

$$= \frac{1}{3} \times \frac{2}{5}\left[12 + \frac{294}{25} + \frac{972}{25} + \frac{726}{25} + \frac{2028}{25} + 27\right]$$

$$\frac{2}{15}\left[12 + \frac{294}{25} + \frac{972}{25} + \frac{726}{25} + \frac{2028}{25} + 27\right] = \frac{2}{15} \times 199.8$$

$$= 26.64$$

The absolute error is given as:

E = value of the integral function-value of the midpoint

$$E = |26 - 26.64| \approx -0.64$$

The relative error is calculated as:

$$R = \frac{Absolute\ error\ (E)}{Exact\ value} = \frac{0.36}{26}$$

$$R \approx -0.0246 \approx -2.46\%$$

Based on the function that was considered, it is observed that Simpson's rule was more accurate than trapezoid rule. The midpoint rule was less accurate than both. This scenario depends on the function that was considered.

1.4 Computational application of numerical integration

In this section, codes are written using the MATLAB to estimate the approximation of each rules computationally. The Simson's rule for solving the problem posed in the previous section is given in Code7.1. The code is perfect for all number of segments that is even.

Code7.1

```
% Create an unknown variable 'x'
    1.  syms x
% Lower Limit
    2.  x0 = 1;
% Upper Limit
    3.  xn = 3;
% Number of Segments where n must be even at all times
    4.  n = 6;
% Declare the function
    5.  func = 3*x^2;
% inline creates a function of string containing in func
    6.  f = inline(func);
% h is the segment size
    7.  h = (xn - x0)/n;
% Create the odd and even terms respectively
    8.  Odd = 0;
    9.  Even = 0;
% Create the non-odd and non-even terms respectively
    10. si=0;
```

```
11. so=0;
12. for i = 1:1:n-1
```

%Ensure that the odd terms are correct
```
13. if rem(i,2)==1
14. xi=a+(i*h);
15. Odd=Odd+f(xi); %sum of all odd terms
16. else
```
%Discard the non-Odd terms
```
17. si=si+f(xi);%sum of others terms
18. end
19. end
20. for i = 2:1:n-2
```
%Ensure that the even terms are correct
```
21. if rem(i,2)==0
22. xi=a+(i*h);
23. Even=Even+f(xi); %sum of all even terms
24. else
```
%Discard the non-Even terms
```
25. so=so+f(xi);%sum of others terms
26. end
27. end
```
% The summation of the terms
```
28. Answer = (h/3)*(f(x0)+4*Odd+2*Even+f(xn));
```
%Display the results
```
29. disp(Answer);
```

Due to possibility in assessing the impact of the segment number within a certain range, 2D plots shows the convergence of the approximations. The code is displayed in Code7.2

Code7.2

```
% Create an unknown variable 'x'
1.  syms x
% Lower Limit
2.  x0 = 1;
% Upper Limit
3.  xn = 3;
% Number of Segments where n within a specific range
4.  n = linspace(9,3);
% Declare the function
5.  func = 3*x^2;
% inline creates a function of string containing in func
6.  f = inline(func);
% h is the segment size
7.  h = (xn - x0)./n;
% Create the odd and even terms respectively
8.  Odd = 0;
9.  Even = 0;
```

% Create the non-odd and non-even terms respectively
```
10. si=0;
11. so=0;
12. for i = 1:1:n-1
```
%Ensure that the odd terms are correct
```
13. if rem(i,2)==1
14. xi=a+(i*h);
15. Odd=Odd+f(xi); %sum of all odd terms
16. else
```
%Discard the non-Odd terms
```
17. si=si+f(xi);%sum of others terms
18. end
19. end
20. for i = 2:1:n-2
```
%Ensure that the even terms are correct
```
21. if rem(i,2)==0
22. xi=a+(i*h);
23. Even=Even+f(xi); %sum of all even terms
24. else
```
%Discard the non-Even terms
```
25. so=so+f(xi);%sum of others terms
26. end
27. end
28.
29. % The summation of the terms
30. J = (h/3).*(f(x0)+4*Odd+2*Even+f(xn));
```
%Display the results
```
31. plot(n,J,'b','linewidth',3);
32. xlabel('Number of segment');
33. ylabel('Approximations');
```
% Increase the font in both axes
```
34. set(gca,'FontSize',18);
```

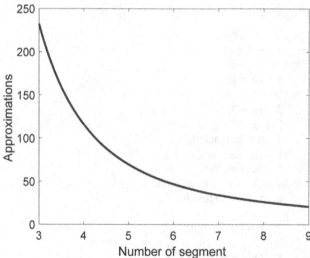

The next assignment is to code the trapezoid rule using the MATLAB to find the approximation. The code is displayed in Code7.3 for any number of segment as displayed below.

Code7.3

```
% Create an unknown variable 'x'
    1.  syms x
% Lower Limit
    2.  x0 = 1;
% Upper Limit
    3.  xn = 3;
% Number of Segments where n must be even at all times
    4.  n = 6;
% Declare the function
    5.  func = 3*x^2;
% inline creates a function of string containing in func
    6.  f = inline(func);
% h is the segment size
    7.  h = (xn - x0)/n;
% Create the odd and even terms respectively
    8.  Term = 0;
    9.  for i = 1:1:n-1
    10. xi=a+(i*h);
    11. Term=Term+f(xi);
    12. end
% The summation of the terms
    13. Answer = (h/2)*(f(x0)+2*Term+f(xn));
%Display the results
    14. disp(Answer);
```

Due to possibility in assessing the impact of the segment number within a certain range, 2D plots shows the convergence of the approximations. The code is displayed in Code7.4 below.

Code 7.4

```
% Create an unknown variable 'x'
1.  syms x
% Lower Limit
2.  x0 = 1;
% Upper Limit
3.  xn = 3;
% Number of Segments where n within a specific range
4.  n = linspace(9,3);
% Declare the function
5.  func = 3*x^2;
% inline creates a function of string containing in func
6.  f = inline(func);
% h is the segment size
7.  h = (xn - x0)./n;
% Create the odd and even terms respectively
8.  Term = 0;
9.  for i = 1:1:n-1
10. xi=a+(i*h);
11. Term=Term+f(xi);
12. end
% The summation of the terms
13. J = (h/2).*(f(x0)+2*Term+f(xn));
%Display the results
14. plot(n,J,'b','linewidth',3);
15. xlabel('Number of segment');
16. ylabel('Approximations');
% Increase the font in both axes
17. set(gca,'FontSize',18);
```

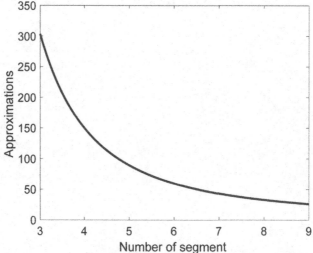

The next assignment is to code the midpoint rule using the MATLAB to find the approximation. The code is displayed in Code7.5 for any number of segments as displayed below.

Code7.5

```
% Create an unknown variable 'x'
1.  syms x
% Lower Limit
2.  x0 = 1;
% Upper Limit
3.  xn = 3;
% Number of Segments where n must be even at all times
4.  n = 4;
% Declare the function
5.  func = 3*x^2;
% inline creates a function of string containing in func
6.  f = inline(func);
% h is the segment size
7.  h = (xn - x0)/n;
% Create the odd and even terms respectively
8.  Term = 0;
9.  for i = 1:1:n-1
10. xg=a+(i*h);
11. xi=(xg+(xg+(i*h)))/2;
12. Term=Term+f(xi);
13. end
% The summation of the terms
14. Answer = h*Term;
%Display the results
15. disp(Answer);
```

Due to possibility in assessing the impact of the segment number within a certain range, 2D plots shows the convergence of the approximations. The code is displayed in Code7.6 below.

Code7.6

```
% Create an unknown variable 'x'
 1.  syms x
% Lower Limit
 2.  x0 = 1;
% Upper Limit
 3.  xn = 3;
% Number of Segments where n within a specific range
 4.  n = linspace(9,3);
% Declare the function
 5.  func = 3*x^2;
% inline creates a function of string containing in func
 6.  f = inline(func);
% h is the segment size
 7.  h = (xn - x0)./n;
% Create the odd and even terms respectively
 8.  Term = 0;
 9.  for i = 1:1:n-1
10.  xg=a+(i*h);
11.  xi=(xg+(xg+(i*h)))/2;
12.  Term=Term+f(xi);
13.  end
% The summation of the terms
14.  J = h.*Term;
%Display the results
15.  plot(n,J,'b','linewidth',3);
16.  xlabel('Number of segment');
17.  ylabel('Approximations');
% Increase the font in both axes
18.  set(gca,'FontSize',18);
```

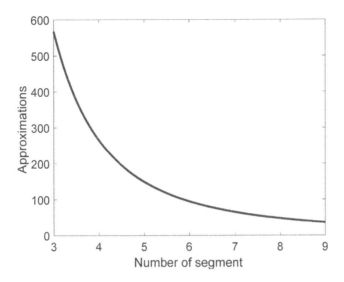

As discussed previously on the accuracy of the numerical integration techniques, the 2D graph is further evidence that the Simpson's rule was more accurate than trapezoid rule. The midpoint rule was less accurate than both. This scenario depends on the function and number of segment that was considered.

Next the application of numerical integration to dataset is demonstrated. The dataset used for this section is presented below. The code for the first dataset is presented in Code7.7. The second code is for the combination of the two datasets.

Code7.7

```
%define the names of all the excel files
1.  data=xlsread('exp.xls');
2.  x = data(:,1);
3.  y = data(:,2);
4.  y1 = data(:,3);
%Find the numerical integration of the data
5.  v = cumtrapz(x, y);
6.  c = cumtrapz(x, y1);
%Plot the dataset
7.  figure
8.  plot(x,v,'-b',x,c,'r','linewidth',3);
9.  ylabel('y');
10. xlabel('x');
11. legend('Integral-data1', 'Integral-data2');
12. set(gca,'FontSize',18)
```

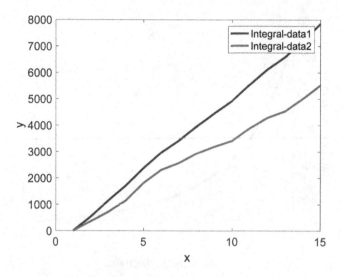

A	B	C
1	467.79	339.25
2	612.47	386.8
3	592.9	323.33
4	542.32	535.35
5	795.11	807.6
6	379.32	175.07
7	521.41	338.95
8	563.21	391.58
9	442.82	137.01
10	502.23	309.87
11	703.11	620.69
12	472.2	186.19
13	466.6	308.43
14	702.72	625.37
15	643.45	401.13

The derivation of the polynomic function of the derivative is presented below and the coefficients displayed in a table. The coefficient table is important to synthesize the boundary conditions in live scenarios as proven by Pochai (2017).

Code7.8

```
%define the names of all the excel files
    1.  data=xlsread('exp.xls');
    2.  x = data(:,1);
    3.  y = data(:,2);
    4.  y1 = data(:,3);
%Find the numerical integral of the data
    5.  v = cumtrapz(x, y);
    6.  c = cumtrapz(x, y1);
%Find the polyfit of function for derivative
    7.  p = polyfit(x,v,14);
    8.  p2 = polyfit(x,c,14);
    9.  y2 = polyval(p,x);
    10. y3 = polyval(p2,x);
%Find the polyfit of function for derivative
    11. p3 = polyfit(x,y,14);
    12. p4 = polyfit(x,y1,14);
%Plot the dataset
    13. figure
    14. plot(x,v,'bo',x,y2,'r-',x,c,'go',x,y3,'m-','linewidth',3)
    15. ylabel('y');
    16. xlabel('x');
    17. legend('Integral-data1','Integral-polyfit1','Integral-data1', 'Integral-polyfit2');
    18. set(gca,'FontSize',18)
```

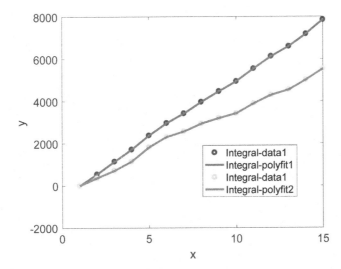

Coefficient table

Coeff_Der_1	Coeff_data1	Coeff_Der_2	Coeff_data2
−7.92E-07	4.67E-06	−1.00E-06	3.74E-06
9.20E-05	−0.0005344	0.00011778	−0.0004325
−0.0048463	0.02777024	−0.0062532	0.02277933
0.15296012	−0.867564	0.19870896	−0.7221609
−3.2239392	18.1653158	−4.2112957	15.3588098
47.8653754	−269.00208	62.7864315	−231.16355
−514.56925	2896.84789	−676.88804	2530.6511
4054.15616	−22964.883	5340.91015	−20389.78
−23425.568	134106.767	−30865.907	120929.826
98313.0705	−571151.14	129406.727	−522465.29
−293358	1735486.47	−385372.42	1607897.33
599502.152	−3620365	785509.374	−3390712.1
−786591.35	4853795.92	−1027869.5	4585963.17
587156.883	−3695638.6	765088.763	−3515485.1
−185181.56	1184553.07	−240619.84	1132287.05

References

Openstax, 2022. Numerical Integration. https://openstax.org/books/calculus-volume-2/pages/3-6-numerical-integration.

Pochai, N., 2017. Unconditional stable numerical techniques for a water-quality model in a non-uniform flow stream. Advances in Difference Equations 2017, 1−13.

Numerical interpolation in environmental research

1. Introduction

Interpolation is a very important tool in environmental modeling as it used for hindcast, nowcast, and forecast. It makes use of dataset (i.e., discrete remote, sampling, or experimental data) to formulate function that could simulate or imitate the trend of the data points with minimal deviations. Interpolation estimates the intermediate value of a function while extrapolation estimates value of a function outside a given range. This process enables the user to make intelligent estimation of data between known data points. The second importance of interpolation in environmental sciences is its ability to simplify complicated functions by sampling data points and interpolating them using a simpler function. The formulation of the interpolation function via the dataset are essential through the use of polynomials. The polynomial route is widely adapted during interpolation because of the ease to evaluate, differentiate, and integrate the function. Also, since interpolation is all about fitting a smooth curve through a given set of points, the description of the smooth curve mathematically can only be achieved via polynomial.

The third importance of interpolation in environmental science is the generating of datasets based on the formulated function. This scenario is very common when there are missing datasets due to measurement errors or abnormalities. The fourth importance of interpolation in environmental sciences is the ease to adopt it for optimization processes. For example, if there are 108 discrete data points that was obtained during a field study and during the analysis of the field dataset, the scientist observed that 111 discrete data points are needed to make critical decision or to reprocess the dataset using machine learning, then interpolation becomes very handy. The fifth importance of interpolation in environmental sciences is its adaptability to formulate high-order quadrature and differentiation rules.

There are different types of interpolation (i.e., piecewise constant interpolation, linear interpolation, polynomial interpolation, and spline interpolation), but the use of any of the type is largely dependent on speed of execution, accuracy, cost, number of data points needed, and user friendliness. For example, linear interpolation works with fewer dataset and it is quick and easy to use. Among its disadvantages is that its solution it is not very precise and data points is not differentiable. Bicubic interpolation in two dimensions, and trilinear interpolation in three dimensions are used

Numerical Methods in Environmental Data Analysis. https://doi.org/10.1016/B978-0-12-818971-9.00005-3

when working on gridded or scattered data. Piecewise constant interpolation on the other hand is preferred due to speed and simplicity when user intend to work on higher-dimensional multivariate interpolation.

The disadvantage of the polynomial interpolation is the tendency that the error is proportional to the distance between the data points to the power n. Most times, its error is enormous; hence, the need to always confirm its deviation while working extensively on interpolation for environmental research. Polynomial interpolation is computationally expensive compared to linear interpolation. Hence the need for users to adapt spline interpolation. Spline interpolation uses low-degree polynomials in each of the intervals, and chooses the polynomial pieces such that they fit smoothly together. When user is working on an infinite number of data points, the Whittaker–Shannon interpolation is recommended. Specific environmental dataset falls within this specification, hence, the need for environmental expert to have hands-on experience on the different types of interpolation. In conclusion, user can construct their own type of interpolation for their work. For example, the rational interpolation is combination of rational functions and Padé approximant, while the trigonometric interpolation is the combination of trigonometric polynomials and Fourier series.

2. Application of interpolation to environmental data

As discussed in the early chapters, experimental or remote measurements are characterized with different types of error such as human error, device calibration, precision of the measuring device etc. Interpolation becomes handy in this case if the data by a common function. There are several ways of finding suitable functions for dataset in environmental research, i.e., interpolation (Deligiorgi and Philippopoulos, 2011), modeling (Emetere, 2016), and learning algorithm (Schaeffer, 2017). In other words, interpolation is a mathematical function that estimates the values of a data where no measured values are available (Baillargeon, 2005). Environmental modeling are mostly time dependent where data are acquired in an equidistant time intervals which may be short or long term. Interpolation becomes handy in such scenario as it applicable in numerical solution of ordinary and partial differential equations, and best fitting problem for finite sets of data (Rohit, 2018).

Interpolation is popularly used for big or large and sparse data sets because it has ability to break the data into smaller subintervals so that the sub-interpolants can be manipulated to fit users' purpose. Roten et al. (2021) worked on satellite observations and emission inventories using modeling processes. The authors wanted to increase the observation density that requires an amount of computational time for large number of spatially distributed, column-based receptors and large carbon dioxide sources. The interpolation method was therefore adopted to reduce resource requirements. The authors reported that the interpolation method gave faster processing time. Constructing a continuous function for data fitting is somewhat complex considering the type of discrete dataset, assumptions of interpolation points, and rates of change or requirements for fitting function.

Deligiorgi and Philippopoulos (2011) worked on environmental air pollution in large city using a special kind of interpolation i.e., spatial interpolation. Spatial interpolation is the procedure of estimating the values of the variable under study at unsampled locations, using point observations within the same region. They reviewed the statistical spatial interpolation methodologies which are used in air pollution modeling. The driving objective for using interpolation was to perform point estimations of pollution fields of densely populated areas in an urban settlement. The Artificial Neural Network (ANN) approach was incorporated into the spatial interpolation and its result was compared to the traditional interpolation techniques. Recall we had previously mention in the preceding section that the concept of interpolation is very flexible for user to combine several techniques with any type of the traditional interpolation technique. In the case of this author, they reported that the combination of ANN and spatial interpolation was found to be specifically promising. However, like any interpolation schemes, the errors observed were from the pollutant of concern and the monitoring network. It is salient to mention at this point that interpolation does not treat erroneous dataset because like computer, the input determines the output.

Ogbozige et al. (2018) worked on water quality assessment of local stream using the weighted interpolation. The weighted interpolation, originally known as the Inverse Distance Weighted (IDW) interpolation is a deterministic spatial interpolation approach to estimate an unknown value at a location using some known values with corresponding weighted values (ESRI, 2015). The driving objective of their work was to investigate the middle stretch of a location river that has been widely research with focus on the upper and lower stretches of the river. Live dataset that bothers on temperature, turbidity, pH, dissolved oxygen, biochemical oxygen demand (BOD), chemical oxygen demand (COD), total nitrogen and total phosphorus were obtained in the middle stretch of the river. However, the reason interpolation was used in the study was to determine the water quality values at unknown locations using IDW interpolation whose formular is displayed below.

$$Z(s_0) = \sum_{i=1}^{n} \lambda_i(s_0) \cdot Z(s_i)$$

Where the weights i.e., $\lambda_i(s_0)$ is given as:

$$\lambda_i(s_0) = \frac{\dfrac{1}{\beta d(s_0, s_1)}}{\displaystyle\sum_{i=0}^{n} \dfrac{1}{\beta d(s_0, s_1)}}; \ \beta > 1$$

λ_i is weights of interpolator, $d(s_0, s_1)$ is the distance from the new point to a known sample point, β is the coefficient used to adjust the weights, and n is the total number of points in the neighborhood analysis. Though certain recommendations were made in the work, the error of the process was not reported, hence it is assumed that errors associated to interpolation technique were reported as part of the discovery. This omission should be avoided when using interpolation for environmental studies because most studies on environment are not read by academic only. Policymakers work largely on environmental report to make informed decisions.

Wong et al. (2004) worked on spatial interpolation methods for estimating air quality data. The work was driven by the objective of assessing the role of infant exposure to ambient air pollutants at pollution sites. Interpolation technique was used in this work to narrow the sampling options and validate experimental values i.e., estimate O_3 and PM10 air concentrations. In this case, block groups to counties were examined. Four different spatial interpolation methods i.e., spatial averaging, nearest neighbor, inverse distance weighting, and kriging was applied to the monitoring data to derive air concentration levels. Though the authors sparingly discussed how errors were mitigated, there were no reported effort of estimating the errors in the dataset before applying the interpolation technique. The authors reported that the different spatial interpolation methods did not produce significantly different estimations where monitor density was relatively low but observed significant differences were monitor density was high. This study shows that using interpolation of large dataset may require additional aiding tools i.e., machine learning tools to mitigate process errors.

In environment research, there is the need to narrow down to specific region of concern during research. The interpolation technique becomes handy in this scenario. For example, Bezyk et al. (2021) wanting to examine the magnitude and distribution of the mixes of the near-ground carbon dioxide (CO_2) components to estimate present and future climate impacts, used anisotropy and interpolation parameters such as neighborhood type and number of sectors angle type, and effective terrain weights. The driving objective for using interpolation was craft optimal sampling strategy to curb excessive CO_2 pollution. Also, the authors wanted to use the interpolation technique to determine CO_2 concentrations in areas where no observations are available. Five interpolation techniques were used in this study. The interpolation techniques include inverse distance weighting (IDW), spline, natural neighbor (NB) interpolation, interpolation based on a triangulated irregular network (TIN), Voronoi polygons, and ordinary kriging. They reported that the IDW and spline interpolation demonstrated higher accuracy when compared to natural neighbor interpolation and Voronoi (Thiessen) polygons. The next sections, three types of numerical interpolation will be discussed in details.

3. Lagrange interpolation

Lagrange interpolation are born from its polynomials and are widely used for polynomial interpolation. This type of interpolation assumes the minimal value of the data point which makes its error value low in practical applications. In extensive work, Lagrange interpolation is susceptible to Runge's phenomenon of large oscillation. This interpolation has found relevance in multidisciplinary research. Most time, it is applied alongside Newton—Cotes method of numerical integration and Shamir's secret sharing scheme in cryptography. The reason why it is used is its flexibility to work on data points that are at equidistant or not at equidistant. Lagrange's interpolation is easy to use and can be manually adapted using spreadsheet (Archer et al., 2018). The introduction of computation has advanced the use of this interpolation to any analysis.

The Lagrange interpolation is used for different purposes in environmental research. Samalerk and Pochai (2018) worked on a developed a one-dimensional model for advection-diffusion-reaction that describes transport and diffusion problems of pollutants over stream or canal. Though the model works on the Saulyev technique with quadratic interpolated initial-boundary conditions, the Lagrange interpolation technique was used to synthesize their boundary conditions as required. Therefore, aside its direct application as will be seen in this section, it has extensive used as demonstrated by these authors.

Sahajanand et al. (2020) worked on occurrence and degree of dissolved contaminants, as well as the rate and direction of contaminant's movement within the groundwater flow system. The objective of the research was to develop Phenol Groundwater Transport (PGWT) equation but to understudy the groundwater flow system. The model was developed using the Lagrangian interpolation function over a nine noded rectangular element as expressed in the equation below.

$$N_{i(C,P)} = N_{i,1}C^2P^2 + N_{i,2}C^2P + N_{i,3}C^2 + N_{i,4}CP^2 + N_{i,5}CP + N_{i,6}C + N_{i,7}P^2 + N_{i,8}P + N_{i,9}$$

where C is the concentration of the phenol, P is the permeability, and N is the coefficient. The research was done for concentration 50 ppm, 75 and 100 ppm under the permeability of 0.32, 0.38 and 0.43 respectively. With polynomials of degree ten, five and three, they were able to show significant similarity between the phenol concentration and permeability of the ground system.

Ge et al. (2018) worked on the water distribution calculation of mobile sprinkler machine. Since the interest of the study was to determine the spray uniformity coefficients using radial water distribution curves were obtained by adopting cubic spline interpolation, Lagrange interpolation, least-squares polynomial fitting, and simplified geometric curves, respectively. In this case, the Lagrange interpolation was used to determine the radial water distribution curves. The author introduced the Runge's phenomenon into the Lagrange interpolation method but lead to experimental noise i.e., sharp oscillations at both ends of the radial water distribution curve and large inconsistencies in the irrigation depth simulations. This work saliently shows that before a theory or method is introduced in the Lagrange interpolation, user must be sure that they understand the principles guiding the use of the Lagrange interpolation. Based on this, the theories of the Lagrange interpolation is explained so readers could know when and how to introduce or incorporate external methods to it during live applications.

The direct formulation of the Lagrange interpolation is done using the polynomial bellow for a given $n+1$ data points.

$$y = f(x) = a_0 + a_1x + a_2x^2 + \ldots + a_nx^n$$

Generally, if the coefficients i.e., a is known, it will be easy to estimate any missing dataset. Hence, $f(x)$ can be described using Lagrange's interpolation. The direct tiers for the derivation of the Lagrange interpolation is demonstrated below. The first is the Lagrange first order interpolation.

Given

$$f(x) = f(x_0) + (x - x_0)\frac{f(x_1) - f(x_0)}{x_1 - x_0}$$

If $f(x_0) = f_0, f(x_1) = f_1$ etc., then

$$f(x) = f_0 + \frac{(x - x_0)}{x_1 - x_0}(f_1 - f_0)$$

If the above is expanded

$$f(x) = f_0 - \frac{(x - x_0)}{x_1 - x_0}f_0 + \frac{(x - x_0)}{x_1 - x_0}f_1$$

$$f(x) = \frac{(x_0 - x_1) - (x - x_0)}{x_1 - x_0}f_0 + \frac{(x - x_0)}{x_1 - x_0}f_1$$

$$f(x) = \frac{(x_1 - x_0) - (x - x_0)}{x_1 - x_0}f_0 + \frac{(x - x_0)}{x_1 - x_0}f_1$$

$$f(x) = \frac{(x_1 - x)}{x_1 - x_0}f_0 + \frac{(x - x_0)}{x_1 - x_0}f_1$$

The expansion of these Lagrange first order interpolation to the nth term gives:

$$y = f(x) = \frac{(x - x_1)(x - x_2)...(x - x_n)}{(x_0 - x_1)(x_0 - x_2)...(x_0 - x_n)}y_0 + \frac{(x - x_0)(x - x_2)...(x - x_n)}{(x_1 - x_0)(x_1 - x_2)...(x_1 - x_n)}y_1 + ...$$
$$+ \frac{(x - x_0)(x - x_1)...(x - x_{n-1})}{(x_n - x_0)(x_n - x_1)...(x_n - x_{n-1})}y_n$$

This formula can be used to do many things. One of the functions not mentioned above is the prediction of more data points. The example explains how it can be done.

Given the table below, find the data point for $y(6)$ using the Lagrange interpolation.

x	12	13	18	23
y	4	6	9	11

The table is modified to assign the n terms of each value.

n	0	1	2	3
x	12	13	18	23
y	4	6	9	11

From the above, it can be seen that the interval is not equidistant. Therefore, the Lagrange interpolation for this table is given as:

$$y(x) = \frac{(x - x_1)(x - x_2)(x - x_3)}{(x_0 - x_1)(x_0 - x_2)(x_0 - x_3)}y_0 + \frac{(x - x_0)(x - x_2)(x - x_3)}{(x_1 - x_0)(x_1 - x_2)(x_1 - x_3)}y_1$$
$$+ \frac{(x - x_0)(x - x_1)(x - x_3)}{(x_2 - x_0)(x_2 - x_1)(x_2 - x_3)}y_2 + \frac{(x - x_0)(x - x_1)(x - x_2)}{(x_3 - x_0)(x_3 - x_1)(x_3 - x_2)}y_3$$

Insert the values from the table into the interpolation

$$y(x) = \frac{(x - 13)(x - 18)(x - 23)}{(12 - 13)(12 - 18)(12 - 23)} \times 4 + \frac{(x - 12)(x - 18)(x - 23)}{(13 - 12)(13 - 18)(13 - 23)} \times 6$$
$$+ \frac{(x - 12)(x - 13)(x - 23)}{(18 - 12)(18 - 13)(18 - 23)} \times 9 + \frac{(x - 12)(x - 13)(x - 18)}{(23 - 12)(23 - 13)(23 - 18)} \times 11$$

$$y(x) = \frac{(x - 13)(x - 18)(x - 23)}{(-1)(-6)(-11)} \times 4 + \frac{(x - 12)(x - 18)(x - 23)}{(1)(-5)(-10)} \times 6$$
$$+ \frac{(x - 12)(x - 13)(x - 23)}{(6)(5)(-5)} \times 9 + \frac{(x - 12)(x - 13)(x - 18)}{(11)(10)(5)} \times 11$$

To find the value of y when $x = 6$ i.e., $y(6)$, the formula becomes

$$y(6) = \frac{(6 - 13)(6 - 18)(6 - 23)}{(-1)(-6)(-11)} \times 4 + \frac{(6 - 12)(6 - 18)(6 - 23)}{(1)(-5)(-10)} \times 6$$
$$+ \frac{(6 - 12)(6 - 13)(6 - 23)}{(6)(5)(-5)} \times 9 + \frac{(6 - 12)(6 - 13)(6 - 18)}{(11)(10)(5)} \times 11$$

$$y(6) = \frac{(-7)(-12)(-17)}{(-1)(-6)(-11)} \times 4 + \frac{(-6)(-12)(-17)}{(1)(-5)(-10)} \times 6 + \frac{(-6)(-7)(-17)}{(6)(5)(-5)} \times 9$$
$$+ \frac{(-6)(-7)(-12)}{(11)(10)(5)} \times 11$$

$$y(6) = 86.55 - 146.88 \times 6 + 42.84 - 10.08 = -27.57$$

The same way lower value of x is found i.e., hindcast, same way a higher value of x is found i.e., forecast. Also, a value between the given data point can also be found i.e., nowcast. This ability makes Lagrange a veritable tool in environmental research. However, the incorporation of an external method needs in-depth understanding as the error may be excessive at the end of the research.

Another derivation of the Lagrange interpolation is via the Vandermonds determinants or from the Newton's divided difference formula. Consider nth degree polynomial where $f(x)$ is approximated at the nth degree polynomial, the Newton's divided difference of nth and $(n+1)^{th}$ is zero i.e., $f(x_0, x_1, ..., x_n) = 0$. Hence, the Newton's divided difference formula is:

$$f(x) = \frac{(x - x_1)...(x - x_n)}{(x_0 - x_1)...(x_0 - x_n)}f_0 + \frac{(x - x_0)...(x - x_n)}{(x_1 - x_0)...(x_1 - x_n)}f_1 + ... + \frac{(x - x_0)...(x - x_{n-1})}{(x_n - x_0)...(x_n - x_{n-1})}y_n$$

From the direct method for the derivation of Lagrange's interpolation, it inferred that Lagrange's and Newton's divided difference approximations are one and the same. However, Lagrange's formula is more convenient to use in computer programming.

4. Newton interpolation

Newton polynomial is also called Newton's divided differences interpolation polynomial because the coefficients of the polynomial are calculated using Newton's divided differences method. The general Newton's interpolation is derived on the assumption that Newton's polynomial $P_n(x_i)$ of n degree is written as:

$$P_n(x_i) = f_i, \quad i = 0, 1, \ldots n - 1$$

$$P_n(x_l) = a_0 + a_1(x - x_0) + a_2(x - x_0)(x - x_1) + a_3(x - x_0)(x - x_1)(x - x_2)$$
$$+ a_4(x - x_0)(x - x_1)(x - x_2)(x - x_3)\ldots + a_n(x - x_0)(x - x_1)(x - x_2)\ldots(x - x_{n-1})$$

Splitting the above according to its data points i.e., $i = 0, 1, \ldots n - 1$.
When $i = 0$

$$f_0 = P_n(x_0) = a_0$$

When $i = 1$

$$f_1 = P_n(x_1) = a_0 + a_1(x - x_0), \quad x = x_1$$

$$f_1 = f_0 + a_1(x - x_0)$$

$$a_1 = \frac{(f_1 - f_0)}{(x_1 - x_0)}$$

When $i = 2$

$$f_2 = a_0 + a_1(x - x_0) + a_2(x - x_0)(x - x_1), \quad x = x_2$$

Substitute a_1, and a_0

$$f_2 = f_0 + \frac{(f_1 - f_0)}{(x_1 - x_0)}(x_2 - x_0) + a_2(x_2 - x_0)(x_2 - x_1)$$

$$a_2(x_2 - x_0)(x_2 - x_1) = f_2 - \left(f_0 + \frac{(f_1 - f_0)}{(x_1 - x_0)}(x_2 - x_0) \right)$$

$$a_2 = \frac{f_2}{(x_2 - x_0)(x_2 - x_1)} - \frac{f_0}{(x_2 - x_0)(x_2 - x_1)} - \frac{(f_1 - f_0)}{(x_1 - x_0)(x_2 - x_1)}$$

$$a_2 = \frac{(f_2 - f_0)}{(x_2 - x_0)(x_2 - x_1)} - \frac{(f_1 - f_0)}{(x_1 - x_0)(x_2 - x_1)}$$

Defining the notation called divided differences

$$f[x_k] = f_k$$

$$f[x_k, x_{k+1}] = \frac{f[x_{k+1}] - f[x_k]}{x_{k+1} - x_k}$$

$$f[x_k, x_{k+1}, \ldots, x_i, x_{i+1}] = \frac{f[x_{k+1}, \ldots, x_{i+1}] - f[x_k, \ldots, x_i]}{x_{i+1} - x_k}$$

So that

$$f_0 = f[x_0]$$

$$a_1 = \frac{(f_1 - f_0)}{(x_1 - x_0)} = f[x_0, x_1]$$

$$a_2 = \frac{(f_2 - f_0)}{(x_2 - x_0)(x_2 - x_1)} - \frac{(f_1 - f_0)}{(x_1 - x_0)(x_2 - x_1)} = \frac{f[x_1, x_2] - f[x_0, x_1]}{(x_2 - x_1)} = f[x_0, x_1, x_2]$$

Hence,

$$a_n = f[x_0, x_1, \ldots, x_n]$$

Newton's divided difference interpolation polynomial can be written as:

$$P_n(x) = f[x_0] + f[x_0, x_1](x - x_0) + f[x_0, x_1, x_2](x - x_0)(x - x_1) + \ldots$$
$$+ f[x_0, x_1, \ldots, x_n](x - x_0)(x - x_1)\ldots(x - x_{n-1})$$

$$P_n(x) = \sum_{k=0}^{n} f[x_0, \ldots x_k] \prod_{i=0}^{k-1} (x - x_i)$$

There are two types of Newton's divided differences interpolation i.e., forward and backward divided difference. In the Newton's forward divided difference, the notation $h = x_{i+1} - x_i$ for each $i = 0, 1, \ldots, k - 1$ and $x = x_0 + sh$. Then the difference $x - x_i$ can be written as $(s - i)h$. Hence, the Newton forward divided difference with the node is $x_0, x_1, \ldots, x_i$ is written as:

$$N_f(x) = [y_0] + [y_0, y_1]sh + \ldots + [y_0, \ldots \ldots y_k]s(s - 1)\ldots(s - k + 1)h^k$$

$$= \sum_{i=0}^{k} s(s - 1)\ldots(s - i + 1)h^i[y_0, \ldots \ldots, y_i]$$

$$= \sum_{i=0}^{k} \binom{s}{i} i! h^i[y_0, \ldots \ldots, y_i]$$

Basically, first forward difference can be written as $y_1 - y_0, y_2 - y_1, y_3 - y_2, \ldots$, $y_k - y_{k-1}$; Or $dy_0, dy_1, dy_2, \ldots, dy_k$. Hence, this can be written in a generic format as:

$$\Delta y_n = y_n - y_{n-1}$$

The forward difference table is given as:

x	y	Δy	Δ²y	Δ³y	Δ⁴y	Δ⁵y
x_0	y_0					
		Δy_0				
$x_1 = x_0 + h$	y_1		$\Delta^2 y_0$			
		Δy_1		$\Delta^3 y_0$		
$x_2 = x_0 + 2h$	y_2		$\Delta^2 y_1$		$\Delta^4 y_0$	
		Δy_2		$\Delta^3 y_1$		$\Delta^5 y_0$
$x_3 = x_0 + 3h$	y_3		$\Delta^2 y_2$		$\Delta^4 y_1$	
		Δy_3		$\Delta^3 y_2$		
$x_4 = x_0 + 4h$	y_4		$\Delta^2 y_3$			
		Δy_4				
$x_5 = x_0 + 5h$	y_5					

The Newton's backward divided difference has a different node i.e., x_i, x_{i-1}, ..., x_0 and it formulated as:

$$N_b(x) = [y_k] + [y_k, y_{k-1}](x - x_k) + \ldots + [y_k, \ldots\ldots\ldots y_0](x - x_k)(x - x_{k-1})\ldots\ldots(x - x_1)$$

The assumption is that the intervals or data points are equally spaced such that $x_0 = x_k + sh$ and $x_i = x_k - (k-i)h$ for $i = 0, 1, \ldots\ldots\ldots, k$ which transforms the above equation to:

$$N_b(x) = [y_k] + [y_k, y_{k-1}]sh + \ldots + [y_k, \ldots\ldots\ldots y_0]s(s+1)\ldots\ldots(s+k-1)h^k$$

$$= \sum_{i=0}^{k} (-1)^i \binom{s}{i} i! h^i [y_k, \ldots\ldots\ldots, y_{k-i}]$$

Basically, first backward difference can be written as y_1-y_0, y_2-y_1, y_3-y_2,... ... ...,y_k-y_{k-1} Or dy_0, dy_1, dy_2, ..., dy_k. Hence, this can be written in a generic format as:

$$\Delta y_n = y_n - y_{n-1}$$

The backward difference table is given as:

x	y	Δy	Δ²y	Δ³y	Δ⁴y	Δ⁵y
x_0	y_0					
		Δy_1				
$x_1 = x_0 + h$	y_1		$\Delta^2 y_2$			
		Δy_2		$\Delta^3 y_3$		
$x_2 = x_0 + 2h$	y_2		$\Delta^2 y_3$		$\Delta^4 y_4$	
		Δy_3		$\Delta^3 y_4$		$\Delta^5 y_5$
$x_3 = x_0 + 3h$	y_3		$\Delta^2 y_4$		$\Delta^4 y_5$	
		Δy_4		$\Delta^3 y_5$		
$x_4 = x_0 + 4h$	y_4		$\Delta^2 y_5$			
		Δy_5				
$x_5 = x_0 + 5h$	y_5					

Let us apply the tables now.

The given table requires the computation of Cos(34).

θ	15	30	45	60
Cos(θ)	0.9659	0.8660	0.7071	0.5

Then the difference table becomes:

x	y	Δy	Δ²y	Δ³y
15	0.9659			
		−0.0999		
30	0.8660		−0.059	
		−0.1589		0.0108
45	0.7071		−0.0482	
		−0.2071		
60	0.5			

Angle 34 is 4 steps ahead of 30.

$$y(34) = 0.8660 - 4h = 0.866 - 4 * (0.0108) = 0.8228$$

5. Spline interpolation

Spline interpolation is a form of interpolation where the interpolant is a special type of piecewise polynomial called a spline. Spline becomes very handy when considering functions that are used in applications requiring data interpolation and/or smoothing. Most graphic user interface software makes use of the spline interpolation for data such as ASCII and image dataset. The interesting part of spline interpolation is its flexibility to fits low-degree polynomials to small subsets of the values. Spline interpolation is often preferred over polynomial interpolation because the interpolation error can be made small even when using low-degree polynomials for the spline. This assertion can be confirmed by the live application of spline to radial water distribution curves presented by Ge et al. (2018) in our previous discussion. In practical terms when working on high-degree polynomial, environmental researcher is advised to use the spline interpolation as it will minimize oscillation between data points. This oscillation anomalies between point is common with the Runge kuta interpolation. Hence, combining the any method with any of the interpolation needs a good understanding of the short coming of each interpolation and how to avoid.

Spline functions are found to be finite dimensional in nature, which is the primary reason for their utility in computations and representation. In recent years, splines have attracted attention of both researchers and users who need various approximation tools. There are few special cases of spline interpolation (such as cubic spline interpolation, linear spine interpolation, quadratic spline interpolation, quartic spline interpolation etc.) but the most common is the cubic spline interpolation. It is most used when the Runge phenomenon is significantly noticed in the generated data points. Linear splines are a set of joint line segments, a continuous function with discontinuous first derivative at the data points. Quadratic splines have a continuous first derivative, cubic splines continuous first and second derivatives, and so on. The quartic spline interpolation has higher accuracy than the cubic spline interpolation (Therneau and Grambsch, 2000). The short coming of the quartic spline interpolation is the cumbersomeness in writing the code. Hence, it is complex for a new beginner to adapt for a series of processes.

The linear splines are much simpler than cubic splines and very convenient for constructing blocks for more complex operations (Carpenter, 2004). One of the shortcomings of the Linear spline models is the abrupt change in trend going from one segment to the next, which does not represent what would naturally occur. In such cases, it is advisable to use the quadratic splines as it has been proven to be flexible to join segments of parabolas. (Huang and Stone, 1999). One of the advantages of quadratic spline interpolation method is drawn from its basic properties i.e., continuously differentiable piecewise function that produces a better fit to a continuous function. Cubic splines is very relevant where smooth interpolation becomes essential i.e., computational animation and image scaling. This makes it a very popular tool in environmental research for image processing. Cubic splines are represented by a cubic polynomial within each interval and has continuous first and second derivatives at the data points. Based on the above, the cubic spline interpolation is discussed more in detail.

Consider intervals i.e., $[x_i, x_{i+1}]$ denoted by a function $S(x)$. Spline is defined as:

$$S_i(x) = a_i(x - x_i)^3 + b_i(x - x_i)^2 + c_i(x - x_i) + d_i \quad \text{for} \quad x \in [x_i, x_{i+1}]$$

From the spline above an interval of n is assumed with four coefficients (i.e., a_i, b_i, c_i, d_i), a total of 4n parameters is required to define the spline. This process leads to two conditions that are considered to be a piecewise continuous function.

$$S_i(x_i) = y_i$$
$$S_i(x_{i+1}) = y_{i+1}$$

To make the spline smooth 2n parameter is needed which are continuous at both first and second derivatives.

$$S'_{i-1}(x_i) = S'_i(x_i)$$
$$S''_{i-1}(x_i) = S''_i(x_i)$$

Hence, with 4n coefficients and 4n linear conditions, it easy to formulate the equations that determine them. This process is easier computationally.

Spline interpolation has found tremendous application in environmental research. Chervenkov et al. (2014) reported the need for an alternative approach for computing exposure indices, such as AOT40 of the ground-level ozone. The aim of the research was to seek for a better way for calculating the exposure indices. The cubic spline interpolation became essential in this research for calculating or simulating accumulated effects of ozone concentrations. The spline formular that was used is displayed below:

$$\Delta: a = x_0 < x_1 < \ldots < x_n = b$$

Using the formula given in the spline polynomial, they obtained

$$S_\Delta(Y; x) = M_j \frac{(x_{j+1} - x)^3}{6h_{j+1}} + M_{j+1} \frac{(x - x_j)^3}{6h_{j+1}} A_j (x - x_j) + B_j$$

where the coefficient is given as:

$$B_j = y_j - M_j \frac{h_{j+1}^2}{6}$$

$$A_j = \frac{y_{j+1} - y_j}{h_{j+1}} - \frac{h_{j+1}}{6}\left(M_{j+1} - M_j\right)$$

When the result of the spline interpolation was compared to the classical approach, they observed that the relative difference of values computed reached 15% for relatively short time periods (up to 48 h) and the difference between the two approaches was minimal for relatively long periods.

Li et al. (2019) saw the need to help researchers obtain an approximate form of the original data of Phytoplankton distribution in a Marine environment. Spline interpolation was used in this research because the available observations for the NPZD Type Ecosystem model are usually sparse and uneven. The spline and Cressman interpolations were incorporated into the model. They observed that the distributions can be better inverted with the spline interpolation. Also, the model experiments and results verify the feasibility and effectiveness of spline interpolation. Sekse et al. (2012) saw the need to compare growth differences between 13 *Escherichia coli* strains exposed to various concentrations of growth inhibitor lactoferrin in two dissimilar environments. The spline interpolation was used in this study because the observation was sparse and uneven. The linear spline regression was used with two knots to estimate the slopes of each interval in the bacterial growth curves. The researcher reported that the spline interpolation was in agreement with the experimental measurements.

6. Computational application of interpolation

In this section, the different scenarios that has been used in past research work in environmental research was demonstrated so that reader can develop their own research design from the given cases.

Case 1: Determination of Inherent Characteristics of Lagrange Interpolation

In this section, the characteristics of the Lagrange interpolation is demonstrated. The characteristics is important to enable environmental scientist know when to apply the Lagrange Interpolation. From the graph, the interpolation becomes sinusoidal at the last one-third section of the given interval.

```
   %define the names of all the excel files
1. data=xlsread('exp.xls');
2. x = data(:,1);
3. y = data(:,2);
4. y1 = data(:,3);
   %Compute the Lagrange for both data
5. sum=0;
6. sum2=0;
7. for i=1:lcngth(x)
8. p=1;
9. for j=1:length(x)
10. if j~=i
11. c = poly(x(j))/(x(i)-x(j));
12. p = conv(p,c);
13. end
14. end
15. term = p*y(i);
16. term2 = p*y1(i);
17. sum= sum + term;
18. sum2= sum2 + term2;
19. end
   %Plot the dataset
20. figure
21. plot(x,sum,'b-',x,sum2,'r-','linewidth',3)
22. ylabel('y');
23. xlabel('x');
24. legend('lagrange-data1','lagrange-data2');
25. set(gca,'FontSize',18)
```

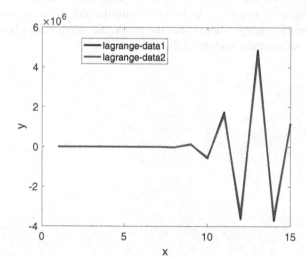

Case 2: Lagrange as finder of polynomial coefficient.

In this section, the added advantage of the Lagrange interpolation was demonstrated. The Lagrange interpolation, the polynomic coefficient of an experimental data, and the coefficient of the first derivative of an experimental data was plotted. It was seen that the Lagrange interpolation is the same as the polynomic coefficient of an experimental data. Hence, if reader wishes to determine the initial condition of a model using Lagrange, the determination of the derivative would no longer be necessary.

```
%define the names of all the excel files
data=xlsread('exp.xls');
x = data(:,1);
y = data(:,2);
y1 = data(:,3);
y2 = data(:,4);
y3 = data(:,5);
%Compute the Lagrange for both data
sum=0;
for i=1:length(x)
    p=1;
    for j=1:length(x)
        if j~=i
            c = poly(x(j))/(x(i)-x(j));
            p = conv(p,c);
        end
    end
    term = p*y(i);
    sum= sum + term,
end
%Plot the dataset
figure
plot(x,sum,'b-',x,y2,'r-',x,y3,'m-','linewidth',3)
ylabel('y');
xlabel('x');
legend('lagrange-data1','Derivative-data1','Coeff-data1');
set(gca,'FontSize',18)
```

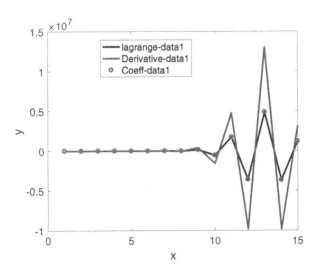

Case 3: Spline interpolation used for tracing experiment data.

In this section, the spline interpolation was used for trendline tracing of an experimental dataset. This technique can be used to know the outliers in the experimental dataset. The outliers are the data points that out of the range of the spline.

```
   %define the names of all the excel files
1. data=xlsread('exp.xls');
2. x = data(:,1);
3. y = data(:,2);
4. y1 = data(:,3);
   %define the querry points
5. xx = 0:1:15;
   %estimate the spline of the data
6. yy = spline(x,y,xx);
   %Plot the graph
7. plot(x,y,'bo',xx,yy,'r-','linewidth',3)
8. ylabel('y');
9. xlabel('x');
10. set(gca,'FontSize',18)
```

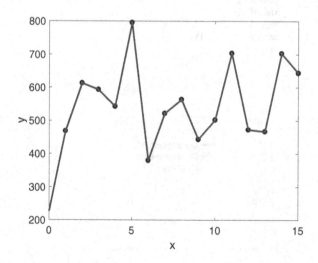

Case 4: Comparing spline interpolation to other interpolations i.e., Piece-wise Cubic Hermite Interpolating Polynomial (PCHIP), and ID interpolation.
In the previous discussion on the application of the interpolation technique, it was established that environmental scientists do a comparative analysis of different interpolations. This technique helps to determine the accurate interpolation technique with very low error. In this section the comparative analysis of different interpolations were demonstrated so reader can build on the existing technique.

```
   %define the names of all the excel files
1.  data=xlsread('exp.xls');
2.  x = data(:,1);
3.  y = data(:,2);
4.  y1 = data(:,3);
   %define the querry points
5.  xq1 = 0:1:15;
   %define the type of interpolations
6.  p = pchip(x,y,xq1);
7.  s = spline(x,y,xq1);
8.  i = interp1(x,y,xq1);
   %Plot the graph
9.  plot(x,y,'o',xq1,p,'-',xq1,s,'-.',xq1,i,'--','linewidth',3)
10. legend('Sample Points','pchip','spline','mkpp','Location','SouthEast');
11. ylabel('y');
12. xlabel('x');
13. set(gca,'FontSize',18)
```

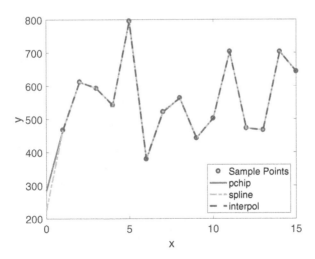

Case 5: Combine other methods (e.g. Bessel) to spline interpolation.

Recall from the previous discussion that the combination of other methods with any type of interpolation is a common practice. However, understanding the method and its compatibility to the experimentation design is very salient for comprehending and making logical scientific guesses. In this section, the Bessel was introduced to the interpolation.

```
%define the names of all the excel files
data=xlsread('exp.xls');
x = data(:,1);
y = data(:,2);
y1 = data(:,3);
%define the querry points
xq1 = 0:1:15;
% introduce the method;
y2 = besselj(1,y);
%define the type of interpolations
p = pchip(x,y2,xq1);
s = spline(x,y2,xq1);
i = interp1(x,y2,xq1);
%Plot the graph
plot(x,y2,'o',xq1,p,'-',xq1,s,'-.',xq1,i,'--','linewidth',3)
legend('Sample Points','pchip','spline','interpol','Location','SouthEast');
ylabel('y');
xlabel('x');
set(gca,'FontSize',18)
```

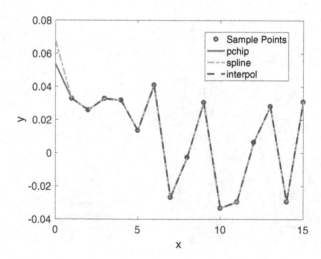

Case 6: Combine other methods (e.g. Beta) to spline interpolation.

In this section, the Beta function was introduced to the interpolation. It can be clearly seen that methods changes the magnitude and nomenclature of the simulation when introduced.

```
%define the names of all the excel files
data=xlsread('exp.xls');
x = data(:,1);
y = data(:,2);
y1 = data(:,3);
%define the querry points
xq1 = 0:1:15;
% introduce the method;
y2 = beta(1,y);
%define the type of interpolations
p = pchip(x,y2,xq1);
s = spline(x,y2,xq1);
i = interp1(x,y2,xq1);
%Plot the graph
plot(x,y2,'o',xq1,p,'-',xq1,s,'-.',xq1,i,'--','linewidth',3)
legend('Sample Points','pchip','spline','interpol','Location','SouthEast');
ylabel('y');
xlabel('x');
set(gca,'FontSize',18)
```

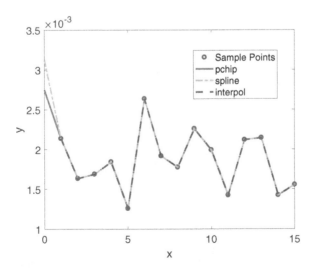

Case 7: Combine other methods (e.g. Lambert W function) to spline interpolation.

In this section, the Lambert W function was introduced to the interpolation. It has now been confirmed that methods changes the magnitude and nomenclature of the simulation when introduced. Based on the above, it is advisable to compare the methods to see which one fits into user's description.

```
   %define the names of all the excel files
1. data=xlsread('exp.xls');
2. x = data(:,1);
3. y = data(:,2);
4. y1 = data(:,3);
   %define the querry points
5. xq1 = 0:1:15;
   % introduce the method;
6. y2 =lambertw(1,y);
   %define the type of interpolations
7. p = pchip(x,y2,xq1);
8. s = spline(x,y2,xq1);
9. i = interp1(x,y2,xq1);
   %Plot the graph
10. plot(x,y2,'o',xq1,p,'-',xq1,s,'-.',xq1,i,'--','linewidth',3)
11. legend('Sample Points','pchip','spline','interpol','Location','SouthEast');
12. ylabel('y');
13. xlabel('x');
14. set(gca,'FontSize',18)
```

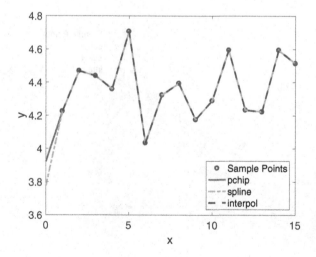

References

Archer, B., Weisstein, Lagrange, E.W., 2018. Langrange Interpolating Polynomial. Retrieved 21 January 2018 from: http://mathworld.wolfram.com/LagrangeInterpolatingPolynomial.html.

Baillargeon, S., 2005. Kriging Review of the Theory and Application to the Interpolation of Precipitation Data. Thesis University of Laval (Quebec), 137 pp.

Bezyk, Y., Sówka, I., Górka, M., Blachowski, J., 2021. GIS-based approach to spatio-temporal interpolation of atmospheric CO_2 concentrations in limited monitoring dataset. Atmosphere 12 (3), 384. https://doi.org/10.3390/atmos12030384.

Carpenter, K.H., 2004. An Introduction to Interpolation and Splines. EECE KSU, pp. 1–8.

Chervenkov, H., Dimov, I., Zlatev, Z., 2014. Spline interpolation for modelling of accumulated effects of ozone. Int. J. Environ. Pollut. 54 (1), 17–31.

Deligiorgi, D., Philippopoulos, K., 2011. Spatial interpolation methodologies in urban air pollution modeling: application for the greater area of metropolitan Athens, Greece. In: Advanced Air Pollution. IntechOpen. https://doi.org/10.5772/17734.

Emetere, M.E., 2016. Numerical Modelling of West Africa Regional Scale Aerosol Dispersion, pp. 65–289. A doctoral thesis submitted to Covenant University, Nigeria.

ESRI, 2015. ArcGIS 10.5: Using ArcGIS Spatial Analyst. Software User Guide ESRI, USA.

Ge, M.S., Wu, P.T., Zhu, D.L., Zhang, L., 2018. Application of different curve interpolation and fitting methods in water distribution calculation of mobile sprinkler machine. Biosyst. Eng. 174, 316–328.

Huang, Z.J., Stone, J.C., 1999. Extended Linear Modeling with Splines. Numerical Mathematics and Computing. seventh edition 250-260, pp. 1–20.

Li, X., Zheng, Q., Lv, X., 2019. Application of the spline interpolation in simulating the distribution of Phytoplankton in a marine NPZD type Ecosystem model. Int. J. Environ. Res. Publ. Health 16 (15), 2664. https://doi.org/10.3390/ijerph16152664.

Ogbozige, F.J., Adie, D.B., Abubakar, U.A., 2018. Water quality assessment and mapping using inverse distance weighted interpolation: a case of river Kaduna, Nigeria. Nigerian J. Technol. 37 (1), 249–261. https://doi.org/10.4314/njt.v37i1.33.

Rohit, R., 2018. Large data analysis via interpolation of functions: interpolating polynomials vs artificial neural networks. Am. J. Intell. Syst. 8 (1), 6–11. https://doi.org/10.5923/j.ajis.20180801.02.

Roten, D., Wu, D., Fasoli, B., Oda, T., Lin, J.C., 2021. An interpolation method to reduce the computational time in the Stochastic Lagrangian particle dispersion modeling of spatially dense XCO2 retrievals. Earth Space Sci. 8. https://doi.org/10.1029/2020EA001343.

Sahajanand, K., Meenal, M., Prasad, G., 2020. Flow and transport of Phenol in groundwater using PGWT equation. Heliyon 6 (2), e03413. https://doi.org/10.1016/j.heliyon.2020.e03413.

Samalerk, P., Pochai, N., 2018. Numerical simulation of a one-dimensional water-quality model in a stream using a saulyev technique with quadratic interpolated initial-boundary conditions. Abstr. Appl. Anal. https://doi.org/10.1155/2018/1926519.

Schaeffer, H., 2017. Learning partial differential equations via data discovery and sparse optimization. Proc. R. Soc. A 473, 20160446. https://doi.org/10.1098/rspa.2016.0446.

Sekse, C., Bohlin, J., Skjerve, E., et al., 2012. Growth comparison of several *Escherichia coli* strains exposed to various concentrations of lactoferrin using linear spline regression. Microb. Inf. Exp. 2, 5. https://doi.org/10.1186/2042-5783-2-5.

Therneau, T.M., Grambsch, P.M., 2000. Modeling Survival Data (Extending the Cox Model). Springer, New York, USA, pp. 100–120.

Wong, D., Yuan, L., Perlin, S., 2004. Comparison of spatial interpolation methods for the estimation of air quality data. J. Expo. Sci. Environ. Epidemiol. 14, 404–415. https://doi.org/10.1038/sj.jea.7500338.

Environmental/ atmospheric numerical models formulations: model review

1. Introduction

In this chapter, various environmental models were discussed to enable readers appreciate the origin of its formulation and the current state of use in the scientific community. This section would help readers see how they can craft their own environmental models.

1.1 Global forecast system

Global Forecast System (GFS) was developed by the National Organization for the Atmosphere in America. The model is freely available for the public view and usage. It can be used to generate data for dozens of atmospheric and land-soil variables, including temperatures, winds, precipitation, soil moisture, and atmospheric ozone concentration (NOAA, 2022). The model originates from the Euler equation given by:

$$\frac{\partial \rho}{\partial t} + \nabla \cdot (\rho \mathbf{u}) = 0$$

$$\frac{\partial (\rho \mathbf{u})}{\partial t} + \nabla \cdot (\rho \mathbf{u} \otimes \mathbf{u}) + \nabla p = -\rho g - 2\rho(\omega \times \mathbf{u}) + P \qquad (9.1)$$

$$\frac{\partial (\rho \theta)}{\partial t} + \nabla \cdot (\rho \theta \mathbf{u}) = Q$$

where q is the density, p is the pressure, u is the velocity vector, g is the gravitational force, x is the vector representing the Earth rotation, h is the potential temperature, that is related to pressure, p, and temperature, T, via $\theta = T/\pi$, with $\pi = (p/p_0)^{R/c_p}$ being the Exner pressure and p_0, R, and c_p being a reference pressure, the gas constant, and the heat capacity at constant pressure, respectively. Equation above needs to be complemented by an equation of state, given as:

$$p = p_0 \left(\frac{\rho R \theta}{p0}\right)^{c_p/c_v} \qquad (9.2)$$

Numerical Methods in Environmental Data Analysis. https://doi.org/10.1016/B978-0-12-818971-9.00006-5

where c_v is the heat capacity at constant volume. In addition, the terms P and Q represents the physical parametrization for the momentum and energy equation, respectively.

Mukhopadhyay et al. (2019) used this model to examine the performance of a very high-resolution global forecast system model at 12.5 km over the monsoon weather in India. The intention of the research is to validate the lead time of monsoon forcast over India. This occurrence can be verified when the mean monsoon rainfall for the season and individual months indicates a tendency for wet bias over the land region. The researcher reported that the model was moderate at capturing the spatio-temporal variability of the monsoon rain; however, there are imperfections in the model to forecast under higher rain rate with a longer lead time. From the understanding of the Euler, it is easier to imagine where and how model that works on that framework perform at certain conditions.

Yue et al. (2022) also worked on the model estimate medium-range precipitation forecasts in the Nile river basin. The work is targeted toward understanding the importance of the Nile to agriculture, hydropower, and weather-related disease outbreaks. Salient shortcoming of the model was reported to be due to large overestimation bias in watersheds located in wet climate regimes in the northern hemispheres and lower ability of the model to capture the temporal dynamics of watershed-averaged rainfall that have smaller watershed areas. Though they proposed the simple climatological bias correction of the NASA's Integrated Multisatellite Retrievals (IMERG) reduced the biases. Hence, to be able to work perfectly using the model, it is recommended that a post analysis using other numerical tools is crucial to reduce errors.

This observation is noticed in other authors work (Patel et al., 2021).

1.2 NOGAPS-ALPHA model

NOGAPS-ALPHA model was developed by the Navy Operational Global Atmospheric Prediction System NOGAPS. Its development and operation is a joint activity of the Naval Research Laboratory NRL and the Navy's Fleet Numerical Meteorology and Oceanography Center FNMOC (Eckermann et al., 2004). This model is used for forecasting photochemistry parameters in the atmosphere using an integrated system that includes data quality control, tropical cyclone bog using, operational data assimilation, balanced initialization, and a global forecast model. Unlike the global forecast model, this model is not available for the public as it is used by defense and civilian users.

The NOGAPS spectral forecast model is a primitive equation model formulated in spherical coordinates in the horizontal and a hybrid vertical coordinate similar to that described by Simmons and Striifing (1981). The horizontal coordinates are the longitude, λ, and the latitude, μ. The vertical coordinate is normalized pressure represented by the variable 77 which ranges from 0.0 at the model top to 1.0 at the surface. If p_{top} is the pressure at the top of the model atmosphere and p_s is the terrain pressure then the pressure p is a function of η given by:

$$p = A(\eta) + B(\eta)\pi \tag{9.3}$$

where $\pi = p_s - p_{top}$. The functions $A(\eta)$ and $B(\eta)$ are any two functions defined on the interval 0.0 to 1.0 with the boundary conditions:

$$\left. \begin{array}{l} A(0) = p_{top} \\ A(1) = p_{top} \\ B(0) = 0.0 \\ B(1) = 1.0 \end{array} \right\} \tag{9.4}$$

where p_{top} is the model's top pressure, p_S the surface pressure and $\pi = p_S - p_{top}$.

The continuity equation for the conservation of mass in this coordinate system is

$$\frac{\partial}{\partial t}\left(\frac{\partial p}{\partial \eta}\right) + \nabla \cdot \left(\mathbf{u}\frac{\partial p}{\partial \eta}\right) + \frac{\partial}{\partial \eta}\left(\dot{\eta}\frac{\partial p}{\partial \eta}\right) = 0 \tag{9.5}$$

where u is the horizontal velocity vector. Integrating (3) with the top and bottom boundary conditions

$$\dot{\eta}(0) = \dot{\eta}(1) = 0 \tag{9.6}$$

yields the π tendency equation

$$\frac{\partial \pi}{\partial t} = -\int_0^1 \nabla \cdot \left(\mathbf{u}\frac{\partial p}{\partial \eta}\right)d\eta = -\int_{p_{top}}^{p_S} \nabla \cdot (\mathbf{u}dp) \tag{9.7}$$

where dp a function of $\lambda, \phi,$ and η. We obtain the vertical motion equation by integrating (3) from 0 to η and substituting for the $\partial \pi / \partial t$ term with the right hand side of (5), yielding

$$\left[\dot{\eta}\frac{\partial p}{\partial \eta}\right](\eta) = B(\eta)\int_0^1 \nabla \cdot \left(\mathbf{u}\frac{\partial p}{\partial \eta}\right)d\eta - \int_0^\eta \nabla \cdot \left(\mathbf{u}\frac{\partial p}{\partial \eta}\right)d\eta \tag{9.8}$$

The thermodynamic energy equation in terms of potential temperature θ is

$$\frac{\partial \theta}{\partial t} = -\frac{u}{a\cos\phi}\frac{\partial \theta}{\partial \lambda} - \frac{v}{a}\frac{\partial \theta}{\partial \phi} - \left[\dot{\eta}\frac{\partial p}{\partial \eta}\right]\frac{\partial \theta}{\partial p} + Q_\theta \tag{9.9}$$

where a is the Earth's radius, u and v are the zonal and meridional wind components, respectively, and Q_θ is the diabatic forcing due to radiation, latent heat release processes, horizontal diffusion, and vertical mixing.

The moisture conservation equation is

$$\frac{\partial q}{\partial t} = -\frac{u}{a\cos\phi}\frac{\partial q}{\partial \lambda} - \frac{v}{a}\frac{\partial q}{\partial \phi} - \left[\dot{\eta}\frac{\partial p}{\partial \eta}\right]\frac{\partial q}{\partial p} + Q_q \tag{9.10}$$

where the forcing term Q_q is due to condensation/evaporation processes and turbulent and cumulus vertical mixing. We write the hydrostatic equation in the form

$$\frac{\partial \phi}{\partial P} = -c_p\theta \tag{9.11}$$

where ϕ is the geopotential and P is the Exner function

$$P = \left(\frac{p}{p_0}\right)^\kappa \tag{9.12}$$

In the above, equation, p_0 is 1000 mb and $\kappa = R/c_p$ is the ratio of the gas constant R to the heat capacity c_p.

The dependent variables describing the motion field in NOGAPS are the vorticity ζ and divergence D. This choice is a routine one for global spectral models because as scalar quantities ζ and D are easily expandable in terms of spherical harmonics. To represent the relationship between ζ, D, and the horizontal wind components it is convenient to define the operator $\alpha(g, h)$, which operates on any two functions g and h, as:

$$\alpha(g, h) = \frac{1}{1-\mu^2}\frac{\partial g}{\partial \lambda} + \frac{\partial h}{\partial \mu} \tag{9.13}$$

where $\mu = \sin\phi$. The vorticity and the divergence are expressed as

$$\zeta = \alpha(V, -U) \tag{9.14}$$

and

$$D = \alpha(U, V) \tag{9.15}$$

where we have defined the scaled wind velocity components

$$U = u\frac{\cos\phi}{a} \tag{9.16}$$

and

$$V = v\frac{\cos\phi}{a}$$

Similarly, the tendency equations for vorticity and divergence are

$$\frac{\partial \zeta}{\partial t} = -\alpha(G, H)$$

and

$$\frac{\partial D}{\partial t} = \alpha(H, -G) - \nabla^2(\phi + I)$$

where the functions G, H, and I are:

$$G = U(\zeta + f) + \left[\dot\eta\frac{\partial p}{\partial \eta}\right]\left(\frac{\partial V}{\partial p}\right) + (1-\mu^2)\frac{c_p}{a^2}\theta\left(\frac{\partial P}{\partial \pi}\right)\left(\frac{\partial \pi}{\partial \mu}\right) - Q_v\frac{\cos\phi}{a}$$

$$H = V(\zeta + f) - \left[\dot\eta\frac{\partial p}{\partial \eta}\right]\left(\frac{\partial U}{\partial p}\right) - \frac{c_p}{a^2}\theta\left(\frac{\partial P}{\partial \pi}\right)\left(\frac{\partial \pi}{\partial \lambda}\right) + Q_u\frac{\cos\phi}{a} \tag{9.17}$$

$$I = \frac{a^2}{1-\mu^2}\left(\frac{U^2 + V^2}{2}\right)$$

and f is the coriolis parameter and Q_u and Q_v are diabatic forcing terms due to surface friction and vertical mixing of momentum.

Many researchers had used this model for several environmental research work (McCormack et al., 2004; Eckermann et al., 2006; Coy et al., 2007).

1.3 Global Environmental Multiscale Model (GEM)

The Global Environmental Multiscale Model (GEM) was developed by the meteorological service of Canada. It is an integrated forecasting and data assimilation system that works on the framework of Global Forecast System and ECMWF's Integrated Forecast System (IFS). The GEM model equations originate from the Euler equations. With the traditional shallow atmosphere approximation, the system of equations in a spherical coordinate (λ, ϕ, r) can be expressed as follows:

$$\frac{du}{dt} - \left(f + \frac{\tan\phi}{a}u\right)v + \frac{1}{\rho}\frac{\partial p}{\partial x} = \left(\frac{du}{dt}\right)_{phys}$$

$$\frac{dv}{dt} + \left(f + \frac{\tan\phi}{a}u\right)u + \frac{1}{\rho}\frac{\partial p}{\partial y} = \left(\frac{dv}{dt}\right)_{phys}$$

$$\frac{dw}{dt} + \frac{1}{\rho}\frac{\partial p}{\partial z} + g = \left(\frac{dw}{dt}\right)_{phys}$$

$$\frac{d\ln\rho}{dt} + \frac{\partial u}{\partial x} + \frac{\partial v}{\partial y} + \frac{\partial w}{\partial z} - \frac{\tan\phi}{a}v = \left(\frac{d\ln\rho}{dt}\right)_{phys}$$

$$\frac{d\ln T}{dt} - \kappa\frac{d\ln p}{dt} = \left(\frac{d\ln T}{dt}\right)_{phys},$$

(9.18)

(9.19)

where Eqs. (9.1)–(9.5) govern the evolutions of the u, y, and w components of velocity, mass, and energy, respectively. The spatial coordinates in the above equations are denoted by (x, y, z), which are related to the spherical coordinate (λ, ϕ, r) through the differential relations given by

$$dx = a\cos\phi d\lambda, \quad dy = ad\phi, \quad dz = dr$$

such that u, y, and w are the physical wind components. In Eq. (9.19), (a) denotes Earth's radius. The Lagrangian derivative in this case can be expressed as

$$\frac{d}{dt} = \frac{\partial}{\partial t} + u\frac{\partial}{\partial x} + v\frac{\partial}{\partial y} + w\frac{\partial}{\partial z}$$

(9.20)

In addition to the four independent variables (x, y, z, t), the system of five Eqs. (9.18)–(9.20) involves six dependent variables, namely, the velocity components u, y, and w, the temperature T, the pressure p, and the density ρ.

Scientists have used this model severally for environmental research (Rivest et al., 1994; Côté et al., 1998; Yeh et al., 2002).

1.4 European Center for Medium Range Weather Forecasts

European Center for Medium Range Weather Forecasts (ECMWF) was established for the development and operation of global models and data-assimilation systems for the dynamics, thermodynamics, and composition of the Earth's fluid envelope and interacting parts of the Earth-system, with a view to preparing forecasts by means of numerical methods; providing initial conditions for the forecasts; and contributing to monitoring the relevant parts of the Earth-system. Also, their mandate includes carrying out scientific and technical research directed toward improving the quality of these forecasts; and the collection and storage of appropriate data (United Nations, 2022).

The evolution equations used in the semi-Lagrangian version of the ECMWF forecast model are (Ritchie, 1997):

Momentum equation

$$\frac{d\vec{V}_H}{dt} + f\vec{k} \times \vec{V}_H + \nabla_H \Phi + R_d T_V \nabla_H \ln p = P_V + K_V \tag{9.21}$$

Thermodynamic equation

$$\frac{dT}{dt} - \frac{\kappa T_V \omega}{(1 + (\delta - 1)Q)p} = P_T + K_T \tag{9.22}$$

Continuity equation

$$\frac{\partial}{\partial t}\left(\frac{\partial p}{\partial \eta}\right) + \nabla_H \cdot \left(\vec{V}_H \frac{\partial p}{\partial \eta}\right) + \frac{\partial}{\partial \eta}\left(\dot{\eta}\frac{\partial p}{\partial \eta}\right) = 0 \tag{9.23}$$

Humidity equation

$$\frac{dQ}{dt} = P_Q \tag{9.24}$$

and Hydrostatic equation

$$\nabla_H \Phi = \nabla_H \Phi_S + \nabla_H \int_1^\eta \left(-R_d T_V \frac{\partial}{\partial \eta'}(\ln p)\right)d\eta' \tag{9.25}$$

Here $\vec{V}_H$ denotes the "horizontal" velocity, ∇_H the horizontal gradient operator, f the Coriolis parameter, T_V virtual temperature, Φ_S surface geopotential, η the (pressure-based) hybrid vertical coordinate (Simmons and Strüffing, 1981), ω the "pressure" vertical velocity (dp/dt), $\kappa = R_d/C_{pd}$, $\delta = C_{pv}/C_{pd}$, and P and K the physical parametrization and the horizontal diffusion contributions, respectively. All other notations follow standard conventions.

The hybrid vertical coordinate $\eta(p, p_s)$ is a monotonic function of pressure and also depends on surface pressure such that $\eta = 1$ for $p = p_s$ and $\eta = 0$ for $p = 0$. Pressure as a function of η is given by

$$p(\eta) = A(\eta) + B(\eta) \cdot p_s$$

where the functions $A(\eta)$ and $B(\eta)$ have to fulfill the conditions

$$A(1) = 0, \quad B(1) = 1, \quad A(0) = B(0) = 0$$

which ensure that $p(1) = p_s$ and $p(0) = 0$. An equivalent definition of the vertical coordinate, but one better suited for the construction of the finite-element based discretization for the vertical with no staggering of variables presented here, is given by the derivative form below

$$\frac{\partial p}{\partial \eta} = \frac{dA}{d\eta} + \frac{dB}{d\eta} p_s$$

Together with the conditions

$$\int_0^1 \frac{dA}{d\eta} d\eta = 0, \quad \int_0^1 \frac{dB}{d\eta} d\eta = 1$$

which ensure that the integral of pressure from the top of the atmosphere to the surface yields the surface pressure p_s. By integrating the continuity equation in the vertical subject to the boundary conditions $\eta = 0$ at $\eta = 1$ and at $\eta = 0$, we obtain expressions for the vertical velocity ω, for $\dot{\eta} \equiv \frac{d\eta}{dt}$, and for the rate of change of surface pressure

$$\omega = -\int_0^\eta \left(\frac{\partial p}{\partial \eta'} D + \frac{dR}{d\eta'} \vec{V}_H \cdot \nabla_H p_s \right) d\eta' + B\vec{V}_H \cdot \nabla_H p_s$$

$$\eta \frac{\partial p}{\partial \eta} = -\frac{\partial p}{\partial t} - \int_0^\eta \nabla_H \cdot \left(\vec{V}_H \frac{\partial p}{\partial \eta'} \right) d\eta'$$

$$\frac{\partial}{\partial t} (\ln p_s) = -\frac{1}{p_s} \int_0^1 \nabla_H \cdot \left(\vec{V}_H \frac{\partial p}{\partial \eta} \right) d\eta = -\frac{1}{p_s} \int_0^1 \left(D\frac{\partial p}{\partial \eta} + \frac{dB}{d\eta} \vec{V}_H \cdot \nabla_H p_s \right) d\eta$$

Here D is horizontal divergence. For the semi-Lagrangian advection scheme an expression for the total time derivative of surface pressure is required:

$$\frac{d}{dt}(\ln p_s) = \int_0^1 \left(\frac{dB}{d\eta} \frac{\partial}{\partial t}(\ln p_s) + \frac{dB}{d\eta} \vec{V}_H \cdot \nabla_H \ln p_s \right) d\eta$$

where $\partial(\ln p_s)/(\partial t)$ is given above.

Like other models mentioned above the ECMWF is one of the most used model in atmospheric and environmental research (Forbes and Ahlgrimm, 2014; Agustí-Panareda et al., 2010; Bechtold et al., 2008; Michalakes et al., 1999).

1.5 Unified Model (UKMO)

Unified Model (UKMO) is developed by the United Kingdom Meteorological Office. It is a Numerical Weather Prediction and climate model that is grid-point based rather than wave based. This model has limited accessibility to the public but to meteorological institutes such as Australian Bureau of Meteorology, the Norwegian Meteorological Institute, the New Zealand National Institute of Water and Atmospheric Research, the Australian Commonwealth Scientific and Industrial Research

Organisation, the South African Weather Service, and the National Centre for Medium Range Weather Forecasting.

Like most environmental and atmospheric models that has been discussed, the Unified Model's atmospheric prediction scheme has its working principles on the differential equation i.e., hydrostatic primitive equations. The compressible Euler equations for a dry atmosphere was captured in advection form in terms of the Exner pressure function π and potential temperature θ as

$$\frac{d\log\rho}{dt} + \nabla \cdot \mathbf{v} = 0$$

$$\frac{d\mathbf{v}}{dt} = -c_p\theta\nabla\pi - g\mathbf{k} \tag{9.26}$$

$$\frac{d\theta}{dt} = Q_\theta.$$

Here, the natural (x, y, z) Cartesian coordinates have been used, the velocity field is given by $\mathbf{v} = (u, v, w)$, $\mathbf{k} = (0, 0, 1)$ denotes the vertical direction and

$$\frac{d}{dt} = \frac{\partial}{\partial t} + \mathbf{v} \cdot \nabla$$

is the Lagrangian derivative. Furthermore, if T is the absolute temperature, ρ is the density and p_0 is a reference pressure value, the thermodynamic variables for an ideal gas can be defined as:

$$\pi = \left(\frac{p}{p_0}\right)^\kappa \quad \theta = \frac{T}{\pi} \quad \rho = \frac{p_0\pi^{\frac{c_v}{R}}}{R\theta} \quad \kappa = \frac{R}{c_p}\frac{1}{\kappa} - 1 = \frac{c_v}{R}.$$

Since

$$\log\rho = \frac{c_v}{R}\log\pi - \log\theta + \log\frac{p_0}{R}$$

The fully compressible equations can then be rewritten as

$$\frac{d\log\pi}{dt} + \frac{R}{c_v}\nabla \cdot \mathbf{v} = \frac{R}{c_v}\frac{d\log\theta}{dt} \tag{9.27}$$

$$\frac{d\mathbf{v}}{dt} = -c_p\theta\nabla\pi - g\mathbf{k}$$

$$\frac{d\theta}{dt} = Q_\theta. \tag{9.28}$$

This process is termed the starting point for the proposed discretization algorithm used for this model. The coriolis and turbulent viscosity terms are omitted in the above equation. However, it is re-introduced when performing an advanced calculation.

The adaptability of the united model by researcher and meteorological institute is huge. Leonard et al. (1997) used the Unified model for numerical weather prediction analyses and forecasts for the Antarctic. They reported that the model had its merit and demerit. With minimal emphasis on the Coriolis force and viscosity term, the influence of the strong katabatic winds in the coastal region initiated a strong coastal easterly wind leading to positive errors in the inland of the coast. They also observed that the model looks perfect in January over the Antarctic but becomes worst in July and beyond, i.e., resulting in excess presence of cloud in its imageries. The author claimed that the model has a good representation of the broadscale field of snow accumulation across the Antarctic. Other researchers that worked on the Unified model include but not limited to Bush et al. (2019) who worked on Regional Atmosphere and Land configuration of the Unified model; Brown et al. (2012) who examined the prospect and shortcoming of the model within 25 years of use; Malcolm et al. (2010) worked on the scalability of the model and how it overcome the shortcoming listed in scenarios reported by Leonard et al. (1997); and Strazdins et al. (2011) worked on the tuning of the model to suite certain analysis.

1.6 French global atmospheric forecast model (ARPEGE)

The two main models of Météo-France are AROME and ARPEGE. They do not have the same purposes but they are complementary. ARPEGE is used to forecast the weather at a high scale in space and time while AROME is used to forecast the weather at a finer scale in space and time. ARPEGE was developed by the French Weather Service, Météo-France, and the European Centre for Medium-range Weather Forecasts (ECMWF). This model was originally targeted for French climate modeling community to understudy the significance of the anthropogenic climate impact. Déqué et al. (1994) used three 10-year integration dataset to validate the accuracy of ARPEGE using prescribed monthly mean sea surface temperatures (SST) observed from 1979 until 1988. The results were compared with the observed climatology for the winter (DJF) and summer (JJA) periods that showed that the model is capable of reproducing the observed climatology.

As other numerical prediction model, ARPEGE is based on a set of Navier–Stokes equations that describes the movements of fluids. This equation is written as

Continuity equation:

$$\frac{\partial\left(\rho U_i^T\right)}{\partial x_i} = 0$$

Momentum equations:

$$\frac{\partial}{\partial x_i}\left(\rho U_i^T U_j^T\right) = -\frac{\partial P}{\partial x_j} - \frac{\partial \tilde{\tau}_{ij}^T}{\partial x_i}$$

Molecular momentum transport:

$$\tilde{\tau}_{ij}^T = -\mu\left(\frac{\partial U_j^C}{\partial x_i} + \frac{\partial U_i^C}{\partial x_j}\right) + \frac{2}{3}\delta_{ij}\mu\frac{\partial U_k^C}{\partial x_k} - \dot{m}_i^D U_j^D - \frac{2}{3}\delta_{ij}\dot{m}_k^D U_k^C$$

Self-diffusive transport of mass:

$$\dot{m}_i^D = -\mu \left(\frac{1}{\rho} \frac{\partial \rho}{\partial x_i} \right) = -\mu \left(\frac{1}{P} \frac{\partial P}{\partial x_i} \right)$$

Like the Unified model, the ARPEGE have limited access to the public but have wide adaptability as seen in the work by Barthelet et al. (1994) who used the model to estimate transient CO_2 emission in the coupled model of ARPEGE and OPAICE; Furevik et al. who used the coupled model of ARPEGE and ICOM to evaluate the Bergen climate model.

1.7 Weather Research and Forecasting (WRF)

The Weather Research and Forecasting (WRF) Model was developed cooperatively by NCEP and the meteorological research community. It is referred to as the next-generation mesoscale numerical weather prediction system designed for both atmospheric research and operational forecasting applications. WRF produce simulations based on actual atmospheric conditions (i.e., measurement) or idealized conditions. WRF offers two dynamical solvers for its computation of the atmospheric governing equations, and the variants of the model are known as WRF-ARW (Advanced Research WRF) and WRF-NMM (nonhydrostatic mesoscale model).

WRF model integrates the compressible, nonhydrostatic Navier–Stokes equations, formulated using a terrain-following mass vertical coordinate η. In order to obtain a thickening of vertical levels in correspondence to the orography, where meteorological dynamics are more complicated because of the interaction between the atmosphere and the orography itself, WRF equations are formulated using a vertical coordinate defined as:

$$\eta = \frac{(p_h - p_{ht})}{\mu} \tag{9.29}$$

with

$$\mu = p_{hs} - p_{ht} \tag{9.30}$$

where p_h is the hydrostatic component of the pressure, and p_{hs} and p_{ht} refer to values along the surface and top boundaries, respectively, and $\mu(x, y)$ represents the mass per unit area within the column in the model domain at (x, y). The equations are cast in flux form using variables that have conservation properties, following the philosophy of Ooyama (1990). In general prognostic equations can be cast in conservative form:

$$\frac{d\mu\xi}{dt} = F_{\Xi} \tag{9.31}$$

where d/dt is the total derivative and F_{Ξ} represents sources and sinks related to variable ξ.

Introducing vertical variable η defined by Eq. (9.29), defining

$$\mathbf{V} = \mu\mathbf{v} = (U, V, W), \quad \Omega = \mu\dot{\eta}, \quad \Theta = \mu\theta \tag{9.32}$$

with $\mathbf{v} = (u, v, w)$, $w = \dot{\eta}$ and   redefining   variables   and   coordinates $(x_1, x_2, x_3) \equiv (x, y, z)$ and $(u_1, u_2, u_3) \equiv (u, v, w)$, NavierStokes equations and other fundamental equations can be reformulated as

$$\frac{\partial U}{\partial t} + (\nabla \cdot \mathbf{V} u) - \frac{\partial}{\partial x}(p\phi_\eta) + \frac{\partial}{\partial \eta}(p\phi_x) = F_U \tag{9.33}$$

$$\frac{\partial V}{\partial t} + (\nabla \cdot \mathbf{V} v) - \frac{\partial}{\partial y}(p\phi_\eta) + \frac{\partial}{\partial \eta}(p\phi_y) = F_V \tag{9.34}$$

$$\frac{\partial W}{\partial t} + (\nabla \cdot \mathbf{V} w) - g\left(\frac{\partial p}{\partial \eta} - \mu\right) = F_W \tag{9.35}$$

$$\frac{\partial \Theta}{\partial t} + (\nabla \cdot \mathbf{V} \theta) = F_\Theta \tag{9.36}$$

$$\frac{\partial \mu}{\partial t} + (\nabla \cdot \mathbf{V}) = 0 \tag{9.37}$$

$$\frac{\partial \phi}{\partial t} + \mu^{-1}[(\mathbf{V} \cdot \nabla \phi) - gW] = 0 \tag{9.38}$$

where θ is the potential temperature, p is the pressure and $\phi = gz$ is the geopotential which evolution is described by Eq. (9.38). In this formulation the right hand side terms F_U, F_V, F_W, and F_Θ represent forcing terms arising from not resolved subgrid scale phenomena, spherical projections, and the earth rotation; these terms, as not explicitly resolved, need to be parameterized in order to introduce in Eqs. (9.33)–(9.38) the effects that they produce on explicitly resolved variables. Eqs. (9.33)–(9.38) with the diagnostic relation for the inverse density α

$$\frac{\partial \phi}{\partial \eta} = -\alpha\mu \tag{9.39}$$

and the equation of state:

$$p = p_0 \left(\frac{R_a \theta}{p_0 \alpha}\right)^\gamma \tag{9.40}$$

are the core of WRF model.

In order to explicitly introduce the effect of moisture, Eqs. (9.33)–(9.40) are reformulated retaining the coupling of dry air mass to the prognostic variables, retaining the conservation equation for dry air and additionally introducing a further prognostic equation for the evolution of mixing ratio of each hydrometeor presents in atmosphere.

Redefining the vertical coordinate with respect to the dry-air mass, η can be written as

$$\eta = \frac{(p_{ah} - p_{aht})}{\mu_a} \tag{9.41}$$

where μ_a is the mass of the dry air in the column and p_{ah} and p_{aht} represent the hydrostatic pressure of the dry atmosphere and the hydrostatic pressure at the top of the dry atmosphere; consequently coupled variables are defined as

$$\mathbf{V} = \mu_a \mathbf{v}, \quad \Omega = \mu_a, \quad \dot{\eta} \; \Theta = \mu_a \theta \tag{9.42}$$

and the moist version of Eqs. (9.33)–(9.38) becomes

$$\frac{\partial U}{\partial t} + (\nabla \cdot \mathbf{V} u) + \mu_a \alpha \frac{\partial p}{\partial x} + \frac{\alpha}{\alpha_a} \frac{\partial p}{\partial \eta} \frac{\partial \phi}{\partial x} = F_U \tag{9.43}$$

$$\frac{\partial V}{\partial t} + (\nabla \cdot \mathbf{V} v) + \mu_a \alpha \frac{\partial p}{\partial y} + \frac{\alpha}{\alpha_a} \frac{\partial p}{\partial \eta} \frac{\partial \phi}{\partial y} = F_V \tag{9.44}$$

$$\frac{\partial W}{\partial t} + (\nabla \cdot \mathbf{V} w) - g \left(\frac{\alpha}{\alpha_a} \frac{\partial p}{\partial \eta} - \mu_a \right) = F_W \tag{9.45}$$

$$\frac{\partial \Theta}{\partial t} + (\nabla \cdot \mathbf{V} \theta) = F_\Theta \tag{9.46}$$

$$\frac{\partial \mu_a}{\partial t} + (\nabla \cdot \mathbf{V}) = 0 \tag{9.47}$$

$$\frac{\partial \phi}{\partial t} + \mu^{-1} [(\mathbf{V} \cdot \nabla \phi) - g W] = 0 \tag{9.48}$$

$$\frac{\partial Q_m}{\partial t} + (\nabla \cdot \mathbf{V} q_m) = F_{Q_m} \tag{9.49}$$

with the diagnostic equation for dry inverse density

$$\frac{\partial \phi}{\partial \eta} = - \alpha_a \mu_a \tag{9.50}$$

and the diagnostic relation for the full pressure (vapor plus dry air)

$$p = p_0 \left(\frac{R_a \theta_m}{p_0 \alpha_a} \right)^\gamma \tag{9.51}$$

In Eqs. (9.43)–(9.49), α_a is the inverse density of dry air $1/\rho_a$ and α is the inverse density taking into account the full parcel density $\alpha = \alpha_a (1 + q_v + q_c + q_r + q_i + \cdots)^{-1}$ where q_* are the mixing ratios (mass per mass of dry air) for water vapor, cloud, rain, ice, etc., while Eq. (9.43) is just the equation for the evolution of mixing ratio of hydrometeors.

This model has found relevance in forecasting purposes. For example, Yáñez-Morroni et al. (2018) used the WRF model for precipitation forecasting in an Andean region; Apicella et al. (2021) used this model to verify the predictive capacity of a year-long verification over Italy; Bibiana et al. (2018) used the model to estimate the vertical profile of meteorological data; Raghavan used the model to evaluate the climate parameters in Vietnam.

1.8 Japan Meteorological Agency Nonhydrostatic Model (JMA-NHM)

Japan Meteorological Agency Nonhydrostatic Model (JMA-NHM) was developed by the Numerical Prediction Division (NPD) of the Japan Meteorological Agency (JMA) in partnership with the Meteorological Research Institute (MRI). The model performance for mesoscale NWP has been verified by comparison with a former operational hydrostatic mesoscale model of JMA. JMA-NHM originally started as Ikawa and Saito's model (Ikawa and Saito 1991) and was expanded to simulate mesoscale phenomena.

The governing basic equations of NHM consist of the following flux form equations on the spherical curvilinear orthogonal coordinate:

$$\frac{\partial}{\partial t}\left(\frac{\rho u}{m_2}\right) + \frac{m_1}{m_2}\frac{\partial p'}{\partial x} = -\text{Adv}_1 + R_1 \tag{9.52}$$

$$\frac{\partial}{\partial t}\left(\frac{\rho v}{m_1}\right) + \frac{m_2}{m_1}\frac{\partial p'}{\partial y} = -\text{Adv}_2 + R_2 \tag{9.53}$$

$$\frac{\partial}{\partial t}\left(\frac{\rho w}{m_3}\right) + \frac{1}{m_3}\frac{\partial p'}{\partial z} = -\frac{1}{m_3}\rho' g - \text{Adv}_3 + R_3. \tag{9.54}$$

Here, $u, v,$ and w represent velocity components; and $p, \rho,$ and g represent the pressure, density, and gravity constants, respectively. Subscripts 1, 2, and 3 correspond to components $x, y,$ and z, respectively; and variables with a prime express perturbation from the hydrostatic reference state in NHM, horizontal average of the initial virtual potential temperature field [see (Eqs. 9.51−9.53) in Saito et al., 2001]. The map factors are denoted as m_1 and m_2 for the x and y directions, respectively; while m_3 is not a map factor in the z direction but a variable introduced for definition of momentums. MRI-NPD/NHM assumed the conformal projection; $m_1 = m_2 = m_3 = m$, while m_2 and m_3 were introduced recently by Yamazaki and Saito (2004) to cope with the cylindrical equidistant projection.

In the operational version for mesoscale NWP at JMA, the Lambert conformal projection is employed, and the map factors are given by

$$m_1 = m_2 = m_3 = \left(\frac{\cos\phi}{\cos\phi_1}\right)^{c-1}\left(\frac{1 + \sin\phi_1}{1 + \sin\phi}\right)^c$$

where ϕ is the latitude of the concerned point and $\phi_I = \pi/6$, and c is given by

$$c = \ln\left(\frac{\cos\phi_1}{\cos\phi_2}\right)\bigg/\ln\left\{\frac{\tan\left(\frac{\pi}{4} - \frac{\phi_1}{2}\right)}{\tan\left(\frac{\pi}{4} - \frac{\phi_2}{2}\right)}\right\}$$

When $\phi_2 = \pi/3, c$ is about 0.72. Terms with Adv denote the advection terms and R the residual terms including curvature, Coriolis, and diffusion terms.

In NHM, density is defined by the sum of the masses of moist air and the water substances per unit volume as

$$\rho = \rho_d + \rho_v + \rho_c + \rho_r + \rho_i + \rho_s + \rho_g$$
$$= \rho_a + \rho_c + \rho_r + \rho_i + \rho_s + \rho_g$$

where subscripts $c, r, i, s,$ and g represent the cloud water, rain, cloud ice, snow, and graupel, respectively; ρ_d is the density of dry air; and ρ_v is the density of water vapor. The continuity equation is as follows:

$$\frac{\partial \rho}{\partial t} + m_1 m_2 \left[\frac{\partial}{\partial x} \left(\frac{\rho u}{m_2} \right) + \frac{\partial}{\partial y} \left(\frac{\rho v}{m_1} \right) \right] + \frac{\partial \rho w}{\partial z}$$

$$= \text{prc} + \underline{\rho \text{DIF} \cdot q_v^*}$$

Here, we neglected the divergence term due to curvature of the earth ($= 2\rho w/a$, where a is radius of the earth), following the shallow assumption. In the right-hand side, pro is the fallout of the precipitable water substances defined by;

$$\text{prc} = \frac{\partial}{\partial z} \left(\rho_a V_r q_r + \rho_a V_s q_s + \rho_a V_g q_g \right)$$

where V is the terminal velocity and q is the mixing ratio of the precipitable water substances. Wacker and Herbert (2003) pointed out that unless the fully barycentric velocity is utilized, diffusion flux appears in the continuity equation. The last (underlined) term in the RHS of Eq. (9.7) is the diffusion of water vapor in unit time, which includes sub grid-scale turbulent mixing and computational diffusion. This term has been newly implemented by Saito et al. (2004) to consider surface evaporation of water vapor for total mass conservation as mentioned in Chapter 8.

The thermodynamic equation is given by

$$\frac{d\theta}{dt} = \frac{\partial \theta}{\partial t} + Adv \cdot \theta = \frac{Q}{C_p \pi} + Dif \cdot \theta$$

where θ is the potential temperature, C_p is the specific heat of dry air at constant pressure, and π is the Exner function.

The state equation is given as the diagnostic equation for density as

$$\rho = \frac{p_0}{R\theta_m} \left(\frac{p}{p_0} \right)^{c_v/c_p}$$

where θ_m is the mass-virtual potential temperature (Saito, 1997) defined by

$$\theta_m = \theta(1 + 0.608 q_v) \left(1 - q_c - q_r - q_i - q_s - q_g \right)$$

Other researchers that have worked on this model include Yamada (2003), Saito et al. (2004), Kumagai (2004), Fujita (2003).

1.9 The fifth generation mesoscale model

The fifth generation mesoscale model is a regional mesoscale model used for creating weather forecasts and climate projections. It was developed by National Center for Atmospheric Research and maintained by Penn State University and the National Center for Atmospheric Research. MM5 is globally relocatable, which helps support different latitudes, terrain types, elevations, soil types, etc. The model can be either hydrostatic or non-hydrostatic, depending on the desired outcome. The fact that the model is regional implies that it requires initial conditions and lateral boundary conditions. This means that each boundary (there are four) has initialized wind speeds, temperatures, pressure and moisture fields (PSU/NCAR, 2005).

In terms of terrain following coordinates (x, y, σ), these are the equations for the nonhydrostatic model's basic variables excluding moisture.

Pressure equation is given as:

$$\frac{\partial p'}{\partial t} - \rho_0 g w + \gamma p \nabla \cdot \mathbf{V} = -\mathbf{V} \cdot \nabla p' + \frac{\gamma p}{T}\left(\frac{\dot{Q}}{c_p} + \frac{T_0}{\theta_0} D_\theta\right)$$

Momentum (x-component) equation is given as:

$$\frac{\partial u}{\partial t} + \frac{m}{\rho}\left(\frac{\partial p'}{\partial x} - \frac{\sigma}{p^*}\frac{\partial p^*}{\partial x}\frac{\partial p'}{\partial \sigma}\right) = -\mathbf{V} \cdot \nabla u + v\left(f + u\frac{\partial m}{\partial y} - v\frac{\partial m}{\partial x}\right) - ew\cos\alpha - \frac{uw}{r_{earth}} + D_u$$

Momentum (y-component) equation is given as:

$$\frac{\partial v}{\partial t} + \frac{m}{\rho}\left(\frac{\partial p'}{\partial y} - \frac{\sigma}{p^*}\frac{\partial p^*}{\partial y}\frac{\partial p'}{\partial \sigma}\right) = -\mathbf{V} \cdot \nabla v - u\left(f + u\frac{\partial m}{\partial y} - v\frac{\partial m}{\partial x}\right) + ew\sin\alpha - \frac{vw}{r_{earth}} + D_v$$

Momentum (z-component) equation is given as:

$$\frac{\partial w}{\partial t} - \frac{\rho_0}{\rho}\frac{g}{p^*}\frac{\partial p'}{\partial \sigma} + \frac{g p'}{\gamma p} = -\mathbf{V} \cdot \nabla w + g\frac{p_0}{p}\frac{T'}{T_0} - \frac{g R_d}{c_p}\frac{p'}{p} + e(u\cos\alpha - v\sin\alpha) + \frac{u^2 + v^2}{r_{earth}} + D_w$$

$$(9.55)$$

Thermodynamics equation is given as:

$$\frac{\partial T}{\partial t} = -\mathbf{v} \cdot \nabla T + \frac{1}{\rho c_p}\left(\frac{\partial p'}{\partial t} + \mathbf{v} \cdot \nabla p' - \rho_0 g w\right) + \frac{Q}{c_p} + \frac{T_0}{\theta_0} D_\theta$$

Advection terms can be expanded as

$$\mathbf{V} \cdot \nabla A \equiv mu\frac{\partial A}{\partial x} + mv\frac{\partial A}{\partial y} + \dot{\sigma}\frac{\partial A}{\partial \sigma}$$

where

$$\dot{\sigma} = -\frac{\rho_0 g}{p^*}w - \frac{m\sigma}{p^*}\frac{\partial p^*}{\partial x}u - \frac{m\sigma}{p^*}\frac{\partial p^*}{\partial y}v$$

Divergence term can be expanded as

$$\nabla \cdot \mathbf{V} = m^2\frac{\partial}{\partial x}\left(\frac{u}{m}\right) - \frac{m\sigma}{p^*}\frac{\partial p^*}{\partial x}\frac{\partial u}{\partial \sigma} + m^2\frac{\partial}{\partial y}\left(\frac{v}{m}\right) - \frac{m\sigma}{p^*}\frac{\partial p^*}{\partial y}\frac{\partial v}{\partial \sigma} - \frac{\rho_0 g}{p^*}\frac{\partial w}{\partial \sigma}$$

Most the research that emanated from the MM5 model was mostly for comparison with other models and for few weather predictions (Colle et al., 2003; Peckham et al., 2004; Smyth et al., 2005).

1.10 Advanced Region Prediction System (ARPS)

The Advanced Region Prediction System (ARPS) was developed at the University of Oklahoma. It is a comprehensive multiscale nonhydrostatic simulation and prediction system that can be used for regional scale weather prediction up to the tornadscale simulation and prediction. It is also used for stormscale atmospheric modeling/prediction system, cloud analysis, long and short-wave radiation analysis, Skamarock adaptive grid refinement, Kain-Fritsch cumulus parameterization, ice microphysics, and ensemble prediction component using Scaled Lagged Averaged Forecasting (SLAF) method. It is a complete system that includes a real-time data analysis and assimilation system, the forward prediction model and a post-analysis package.

The ARPS governing equations are first written in a Cartesian coordinate projected onto a plane tangent to or intercepting the earth's surface. Using standard mathematical relations (Haltiner and Williams, 1980) for the transformation from a local Cartesian space on the sphere to map projection space, we obtain the following equations of motion:

$$\dot{u} = -mp_x\rho^{-1} + (f + f_m)v - \tilde{f}w - uwa^{-1} + F_u \tag{9.56}$$

$$\dot{v} = -mp_y\rho^{-1} - (f + f_m)u - vwa^{-1} + F_v \tag{9.57}$$

$$\dot{w} = -p_z\rho^{-1} - g + \tilde{f}u + (u^2 + v^2)a^{-1} + F_w \tag{9.58}$$

In the above and in the equations to follow, the dot operator denotes the total time derivative, e.g., $\dot{u} \equiv du/dt$, and subscripts $t, x, y, z, \xi, \eta,$ and ζ denote partial temporal or spatial derivative, e.g., $u_x \equiv \partial u/\partial x$. In obtaining Eqs. (9.56)–(9.58), there is an assumption that there is no approxima as presented $u_x \equiv \partial u/\partial x$. Hence, the ellipticity of the earth is neglected and the atmosphere is assumed to be thin so that the radius is replaced by the mean earth radius at the sea level, a. Note that the spatial derivative of map factor due to curvature are retained in $f_m \equiv um_y - vm_x + u\tan(\phi)/a$, as are the Coriolis terms due to vertical motion (those involving $\tilde{f}$). Note that for this system, only gravitational, pressure gradient, and frictional forces (F terms) can change kinetic energy. All other terms cancel each other in the kinetic energy equation.

The equations of state for moist air (see Dutton, 1986), mass continuity, heat energy conservation, and conservation of hydrometeor species are, respectively,

$$\rho = p(R_dT)^{-1}[1 - q_v(\gamma + q_v)^{-1}](1 + q_v + q_{li}) \tag{9.59}$$

$$\dot{\rho} = -\rho\left\{m^2\left[(u/m)_x + (v/m)_y\right] + w_z\right\} \tag{9.60}$$

$$\dot{\theta} = \dot{Q}(C_p\pi)^{-1} \tag{9.61}$$

$$\dot{q} = S_q \tag{9.62}$$

Here $\dot{Q}$ denotes heat source, and S_q represents sources due to moist processes.

Some of the notable work on this model have been reported by several researchers not limited to Xue et al. (1995, 2003) and Gao and Stensrud (2012).

1.11 High Resolution Limited Area Model (HIRLAM)

High Resolution Limited Area Model (HIRLAM) was developed by consortia of collaborating National Meteorological Services (NMSs), i.e., Danish Meteorological Institute (DMI), Estonian Meteorological and Hydrological Institute (EMHI), Finnish Meteorological Institute (FMI), Icelandic Meteorological Office (IMO), Lithuanian Hydrological and Meteorological Service (LHMS), Met Éireann (ME), Norwegian Meteorological Institute (MET), Royal Netherlands Meteorological Institute (KNMI), Agencia Estatal de Meteorología (AEMET), formerly INM, and Swedish Meteorological and Hydrological Institute (SMHI). It was targeted to develop and maintain a numerical short-range weather forecasting system for operational use by the participating institutes. Its prime long-term goal is to provide its members with a state-of-the-art operational short and very short-range numerical weather prediction system, and the expertise associated with it.

The forecast model equations are written based on the general pressure and terrain. The vertical coordinate $\eta(p, p_s)$ is written as

$$\eta(0, p_s) = 0 \text{ and } \eta(p_s, p_s) = 1 \tag{9.63}$$

The formulation corresponds to the hybrid system used at ECMWF (Simmons and Burridge, 1981), with some modifications because the continuity equation is integrated upwards in the HIRLAM model. The model is programmed for a spherical coordinate system (λ, θ), but in the formulation and in the code two metric coefficients (h_x, h_y) have been introduced. This is done to prepare the model for any orthogonal coordinate system or map projection with axes (x, y). For a distance $(\delta X, \delta Y)$ on the earth, it yields that

$$\delta X = ah_x\delta_x \text{ and } \delta Y = ah_y\delta y \tag{9.64}$$

In case of spherical rotated coordinates, we have that

$$\delta X = a\cos\theta\delta\lambda \text{ and } \delta Y = a\delta\delta \tag{9.65}$$

The momentum, thermodynamic, and moisture equations are

$$\frac{\partial u}{\partial t} = (f + \xi)v - \dot{\eta}\frac{\partial u}{\partial \eta} - \frac{R_dT_v}{ah_x}\frac{\partial \ln p}{\partial x} - \frac{1}{ah_x}\frac{\partial}{\partial x}(\phi + E) + P_u + K_u \tag{9.66}$$

$$\frac{\partial v}{\partial t} = -(f + \xi)u - \dot{\eta}\frac{\partial v}{\partial \eta} - \frac{R_dT_v}{ah_y}\frac{\partial lnp}{\partial y} - \frac{1}{ah_y}\frac{\partial}{\partial y}(\phi + E) + P_v + K_v \tag{9.67}$$

where

$$\xi = \frac{1}{ah_xh_y}\left(\frac{\partial}{\partial x}(h_yv) - \frac{\partial}{\partial y}(h_xu)\right) \tag{9.68}$$

$$E = \frac{1}{2}\left(u^2 + v^2\right) \tag{9.69}$$

$$\frac{\partial T}{\partial t} = -\frac{u}{ah_x}\frac{\partial T}{\partial x} - \frac{V}{ah_y}\frac{\partial T}{\partial y} - \dot{\eta}\frac{\partial T}{\partial \eta} + \frac{\kappa T_v \omega}{(1 + (\delta - 1)q)p} + P_T + K_T \tag{9.70}$$

$$\frac{\partial q}{\partial t} = -\frac{u}{ah_x}\frac{\partial q}{\partial x} - \frac{V}{ah_y}\frac{\partial q}{\partial y} - \dot{\eta}\frac{\partial q}{\partial \eta} + P_q + K_q \tag{9.71}$$

The terms P_x and K_x represent tendencies from the physical parameterization and horizontal diffusion, respectively. The hydrostatic equation takes the form

$$\frac{\partial \phi}{\partial \eta} = -\frac{R_d T_v}{p}\frac{\partial p}{\partial \eta} \tag{9.72}$$

and the continuity equation is

$$\frac{\partial}{\partial \eta}\frac{\partial p}{\partial t} + \nabla \cdot \left(\vec{v}_h \frac{\partial p}{\partial \eta}\right) + \frac{\partial}{\partial \eta}\left(\eta \frac{\partial p}{\partial \eta}\right) = 0 \tag{9.73}$$

The definition of the divergence operator is

$$\nabla \cdot \vec{V}_h = \frac{1}{ah_x h_y}\left\{\frac{\partial}{\partial x}\left(h_y u\right) + \frac{\partial}{\partial y}\left(h_x v\right)\right\} \tag{9.74}$$

By integrating the continuity equation, using the boundary conditions $\dot{\eta} = 0$ at $\eta = 0$ and $\eta = 1$, we obtain the equation for the surface pressure tendency

$$\frac{\partial p_s}{\partial t} = -\int_0^1 \nabla\left(\vec{v}_h\frac{\partial p}{\partial \eta}\right)d\eta \tag{9.75}$$

the equation for pressure vertical velocity

$$\omega = \frac{\partial p_s}{\partial t} + \int_\eta^1 \nabla \cdot \left(\vec{V}_h \frac{\partial p}{\partial \eta}\right)d\eta + \vec{V}_h \cdot \nabla p \tag{9.76}$$

and the equation for

$$\dot{\eta}\frac{\partial p}{\partial \eta} - \left(1 - \frac{\partial p}{\partial p_s}\right)\frac{\partial p_s}{\partial t} + \int_\eta^1 \nabla \cdot \left(\vec{V}_h\frac{\partial p}{\partial \eta}\right)d\eta \tag{9.77}$$

The HIRLAM has attracted various research studies which focus on model scalability, configuration, and application (Senkova et al., 2007; de Rooy, 2014; Gleeson et al., 2015).

1.12 Global Environmental Multiscale limited area model

The GEM model was developed by Recherche en Prévision Numérique (RPN), Meteorological Research Branch (MRB), and the Canadian Meteorological Centre (CMC) as an operational weather prediction model in the mid-1990s. It was initially used for air quality chemistry and aerosols from global to the meso-gamma scale. However, it has been scaled-up to estimate vertical extension of the modeling domain to include stratospheric chemistry, aerosols, and formation of polar stratospheric clouds. The GEM's operational model, known as the global deterministic prediction system (GDPS), is currently operational for the global data assimilation cycle and medium-range forecasting, the regional data assimilation spin-up cycle and short-range forecasting.

At the present stage of development, the governing equations are taken to be the forced hydrostatic primitive equations. In the absence of forcing, these equations govern the flow of a frictionless adiabatic shallow hydrostatic fluid on a rotating sphere. A terrain-following hybrid coordinate is defined as

$$\eta \equiv \frac{p - p_T}{p_S - p_T} \equiv \frac{p^* - p_T^*}{p_S^* - p_T^*} \tag{9.78}$$

where

$$p^* = p_T^* + (p_S^* - p_T^*)\eta \tag{9.79}$$

p is pressure, p_S and p_T are its respective values at the bottom and top of the atmosphere, and p_S^* and $p_T^* \equiv p_T$ are the respective bottom and top constant pressures of a motionless isothermal ($T^* \equiv$ constant) reference atmosphere. Eq. (9.79) is a direct consequence of the definition (9.78) of the vertical coordinate, and permits the thermodynamic equation to be written in such a way as to ensure the computational stability of the gravitational oscillations. The governing equations in this coordinate system are thus

$$\frac{d\mathbf{V}^H}{dt} + R_d T_v \nabla \ln p + \nabla \phi + f\left(\mathbf{k} \times \mathbf{V}^H\right) = \mathbf{F}^H \tag{9.80}$$

$$\frac{d}{dt} \ln \left|\frac{\partial p}{\partial \eta}\right| + \nabla \cdot \mathbf{V}^H + \frac{\partial \dot{\eta}}{\partial \eta} = 0 \tag{9.81}$$

$$\frac{d}{dt}\left[\ln\left(\frac{T_v}{T^*}\right) - \kappa \ln\left(\frac{p}{p^*}\right)\right] - \kappa \dot{\eta}\frac{d}{d\eta}(\ln p^*) = F^{T_v} \tag{9.82}$$

$$\frac{dq_v}{dt} = F^{q_v} \tag{9.83}$$

$$\frac{\partial \phi}{\partial \eta} = -R_d T_v \frac{\partial \ln p}{\partial \eta} \tag{9.84}$$

where

$$p = \rho R_d T_v \tag{9.85}$$

and

$$\frac{d}{dt} = \frac{\partial}{\partial t} + \mathbf{V}^H \cdot \nabla + \dot{\eta}\frac{\partial}{\partial \eta} \tag{9.86}$$

are the substantive derivative following the fluid. Here, $\mathbf{V}^H$ is horizontal velocity, $\phi \equiv gz$ is the geopotential height, ρ is density, T_v is virtual temperature, $\kappa = R_d/c_{pd}$, R_d is the gas constant for dry air, c_{pd} is the specific heat of dry air at constant pressure, q_v is specific humidity of water vapor, f is the Coriolis parameter, $\mathbf{k}$ is a unit vector in the vertical, g is the vertical acceleration due to gravity, and

$\mathbf{F}^H, F^{T_v}$, and F^{q_v} are parameterized forcing. Eqs. (9.80)–(9.83) are respectively the horizontal momentum, continuity, thermodynamic, and moisture equations, and Eq. (9.85) is the equation of state, taken here to be the ideal gas law. The prognostic vertical momentum equation of the fully-compressible Euler equations has been reduced to the diagnostic hydrostatic Eq. (9.86).

1.13 ALADIN model

The acronym of ALADINs Aire Limitée Adaptation dynamique Développement InterNational. The ALADIN System is a numerical weather prediction (NWP) system developed by the international ALADIN consortium for operational weather forecasting and research purposes. It is based on a code that is shared with the global model IFS of the ECMWF and the ARPEGE model of Météo-France. Today, this system can be used to provide a multitude of high-resolution limited-area model (LAM) configurations. About 100 scientists, from 15 countries are permanently contributing to the progress of ALADIN NWP system (de Rooy, 2014; Gleeson et al., 2015). Today, as a result of the ALADIN project, two regional NWP models (ALADIN and AROME) are run operationally at Hungarian Meteorological Service (OMSZ).

The vertical coordinate the pressure-based hybrid coordinate η following the orography described by Simmons and Burridge (1981). The following relationship holds:

$$\left.\begin{aligned} \eta(0, \pi) &= 0 \\ \eta(\pi, \pi) &= 1 \\ \frac{\partial \eta}{\partial p}(p, \pi) &> 0 \end{aligned}\right\} \tag{9.87}$$

where π is the surface pressure.

The hydrostatic equation reads:

$$\frac{\partial \Phi}{\partial \eta} = -\frac{RT}{p}\frac{\partial p}{\partial \eta} \tag{9.88}$$

which is used as a diagnostic equation in order to compute the geopotential Φ at level p through an integration in the vertical from the lower boundary condition $\Phi(\pi) = \Phi_s$, where Φ_s denotes the surface geopotential (g times the orography).

The evolution of the parameters which define the atmospheric state in the model, horizontal wind $\vec{v}$, temperature T, and specific humidity q_v is governed by the following equations.

Momentum equation

$$\frac{d\vec{v}}{dt} + 2\vec{\Omega} \times \vec{v} + RT\nabla\ln p + \nabla\Phi = -g\frac{\partial \eta}{\partial p}\frac{\partial \vec{F}_{\vec{v}}}{\partial \eta} + \vec{K}_{\vec{v}} \tag{9.89}$$

Thermodynamic equation

$$\frac{dT}{dt} - \kappa T\frac{\omega}{p} = -\frac{g}{c_p}\frac{\partial \eta}{\partial p}\frac{\partial F_h}{\partial \eta} + K_h \tag{9.90}$$

Moisture equation

$$\frac{dq_v}{dt} = -g\frac{\partial \eta}{\partial p}\frac{\partial F_{q_v}}{\partial \eta} + K_{q_v} \tag{9.91}$$

The right-hand side terms of Eqs. (9.88)–(9.90) are the vertical fluxes F provided by the subgrid scale parameterizations and the horizontal diffusion K of momentum, enthalpy and specific humidity. We assume that both dry air and water vapor are perfect gases and we shall now consider the consequences of this assumption. Following Gill (1982), since dry air is composed mostly (99%) of diatomic molecules and water vapor of triatomic molecules one has

$$c_{pa} = \frac{7}{2}R_a \quad c_{pv} = 4R_v$$

where R_a and R_v are respectively the gas constants for dry air and water vapor, and c_{pa} and c_{pv} the specific heats at constant pressure of dry air and water vapor. The atmosphere being a mixture of dry air and water vapor, R and c_p are given by

$$R = R_a q_a + R_v q_v$$
$$c_p = c_{pa} q_a + c_{pv} q_v$$

where q_a is the specific dry air content. Since we assume (von Bezold, 1888) that the liquid and solid phases of water which appear in the atmosphere are immediately removed by precipitation, one has $q_a = 1 - q_v$. In order to have a fully consistent thermodynamical cycle, the saturated partial pressure of water vapor over water, e_s and ice e_g are computed from the integration of the Clausius-Clapeyron relation

$$\frac{1}{e_s}\frac{de_s}{dT} = \frac{L_v}{R_v T^2} \quad \frac{1}{e_g}\frac{de_g}{dT} = \frac{L_s}{R_v T^2}$$

where L_v and L_s are the latent heats of vaporization and sublimation respectively. We have assumed that the specific volumes of the condensed phases of water are negligible and thus set equal to zero. These formulas are accurate in the meteorological range of temperature when the two latent heats are linear functions of temperature

$$L_v(T) = L_v(T_t) + (c_{pv} - c_l)(T - T_t) \quad L_s(T) = L_s(T_t) + (c_{pv} - c_g)(T - T_t)$$

where T_t is the temperature of the triple point of water, c_l and c_g are the specific heats at constant pressure of liquid water and ice. Thus

$$e_s(T) = e_s(T_t)\exp\left\{ \frac{\left[L_v(0)\left(\frac{1}{T_t} - \frac{1}{T}\right) + (c_l - c_{pv})\ln\frac{T_t}{T} \right]}{R_v} \right\} \tag{9.92}$$

$e_g(T)$ is obtained from changing L_v and c_l into L_s and c_g.

Classically, as there is no source term in the continuity equation, a fictitious substitution process is supposed to occur in order to conserve the total mass of the atmosphere. The mass of water which is removed from the atmosphere by precipitation is replaced by an equivalent amount of dry air, the reverse process taking place when water vapor is supplied to the atmosphere by evaporation at the surface of the Earth. As an option we can avoid to perform this approximation; the continuity equation then reads:

$$\frac{\partial}{\partial \eta}\left(\frac{\partial p}{\partial t}\right) + \nabla \cdot \left(\vec{v}\frac{\partial p}{\partial \eta}\right) + \frac{\partial}{\partial \eta}\left(\dot{\eta}\frac{\partial p}{\partial \eta}\right) = -g\frac{\partial F_p}{\partial \eta} \tag{9.93}$$

F_p denotes the mass flux due to precipitation/evaporation. Through vertical integration, one obtains the evolution equation for surface pressure using, $gF_p(1) + gE$ as lower boundary condition for the mass flux $F_p(1)$ denotes the precipitation which fall on the ground and E the evaporation flux at the surface);

$$\frac{\partial \pi}{\partial t} = -\int_0^1 \nabla \cdot \left(\vec{v}\frac{\partial p}{\partial \eta}\right)d\eta - gF_p(1) - gE \tag{9.94}$$

the pressure coordinate vertical velocity;

$$\omega = -\int_0^\eta \nabla \cdot \left(\vec{v}\frac{\partial p}{\partial \eta}\right)d\eta + \vec{v}\cdot\nabla p - gF_p(\eta) \tag{9.95}$$

and the vertical velocity;

$$\dot{\eta}\frac{\partial p}{\partial \eta} = -\frac{\partial p}{\partial t} - \int_0^\eta \nabla \cdot \left(\vec{v}\frac{\partial p}{\partial \eta}\right)d\eta - gF_p(\eta) \tag{9.96}$$

Expressed in terms of the physical components u and v of the wind field, the equations over the sphere Σ of radius a becomes

$$\left.\begin{array}{l}
\frac{\partial u}{\partial t} - v\zeta + v\nabla^u v + u\nabla^u u + \dot{\eta}\frac{\partial u}{\partial \eta} - fv + RT\nabla^u \ln p + \nabla^u\Phi = -g\frac{\partial \eta}{\partial p}\frac{\partial F_{\vec{v}}^u}{\partial \eta} + K_{\vec{v}}^u \\[2ex]
\frac{\partial v}{\partial t} + vD + u\nabla^u v - v\nabla^u u + \frac{u^2+v^2}{a}\tan\theta + \dot{\eta}\frac{\partial v}{\partial \eta} + fu + RT\nabla^v \ln p + \nabla^v\Phi - g\frac{\partial \eta}{\partial p}\frac{\partial F_{\vec{v}}^v}{\partial \eta} + K_{\vec{v}}^v \\[2ex]
\frac{\partial T}{\partial t} + u\nabla^u T + v\nabla^v T + \dot{\eta}\frac{\partial T}{\partial \eta} - \kappa T\frac{\omega}{p} = -\frac{g}{c_p}\frac{\partial \eta}{\partial p}\frac{\partial F_h}{\partial \eta} + K_h \\[2ex]
\frac{\partial q_v}{\partial t} + u\nabla^u q_v + v\nabla^v q_v + \dot{\eta}\frac{\partial q_v}{\partial \eta} = -g\frac{\partial \eta}{\partial p}\frac{\partial F_{q_v}}{\partial \eta} + K_{q_v}
\end{array}\right\} \tag{9.97}$$

denotes the physical component of the gradient of a given field A in the direction. Their expressions are given by

$$\nabla^u A = \frac{1}{a\cos\theta}\frac{\partial A}{\partial \lambda} \tag{9.98}$$

$$\nabla^v A = \frac{1}{a}\frac{\partial A}{\partial \theta} \tag{9.99}$$

1.14 **Eta model**

The Eta model was the operational limited-area hydrostatic model from June 1993 to June 2006 in the United States. The model uses a unique step-coordinate vertical co-ordinate called the eta (hence the name of the model), a modified version of the sigma coordinate. The model has gone through upgrade processes. the introduction of the so-called sloping steps, or discretized shaved cells topography; piecewise-linear finite-volume vertical advection of dynamic variables; vapor and hydrometeor loading in the hydrostatic equation, severe downslope Zonda wind case, and changes aimed at refining the convection schemes available in the Eta (Chen et al., 1997; Chuang and Manikin, 2001; Mesinger et al., 2012).

The hydrostatic vertical eta (eta) coordinate (Mesinger, 1984) is defined as

$$\eta = \frac{p - p_T}{p_S - p_T}\eta_S \tag{9.100}$$

where

$$\eta_S = \frac{p_{rf}(z_S) - p_T}{p_{rf}(0) - p_T} \tag{9.101}$$

The subscripts S and T stand for the surface and the top of model's atmosphere. $p_{rf}(z)$ is a suitably defined reference pressure expressed as a function of height. z_S denotes predefined reference heights of the terrain which may take only discrete values. With such a step-wise formulation of the lower boundary, unlike with the customary terrain following used in the σ system of Phillips (1966), the coordinate surfaces remain nearly horizontal, which eliminates errors of the pressure gradient force close to the steep terrain (Mesinger and Janjić, 1985). The blocking effect of the step-terrain enforces the component of the flow around the terrain, which is underestimated in the σ system and is believed to contribute to a more realistic precipitation forecast (e.g., Mesinger, 1996).

The coordinate lines on the cubic and the octagonal grids are strictly conformal far from the singular points. Close to the singularities, the conformity constraint is broken, and the coordinates become curvilinear. The smoothed versions of these grids introduced in Purser and Rančić (1997), have the larger minimum grid distance, but the area where the orthogonality does not apply is also broader. Therefore, in order to describe correctly flow on the quasi-uniform grids, the governing equations need to be expressed in terms of a general curvilinear coordinate system.

Let (x, y) define a general curvilinear coordinates on the sphere, and let a^1 and a^2 be the base vectors of the coordinate transformation in the direction of x and y axis, respectively.

The metric tensor of transformation, **G**, is defined as

$$\mathbf{G} = G_{ij} = \mathbf{a_i} \cdot \mathbf{a_j}, \quad i,j = 1,2 \tag{9.102}$$

G is the Jacobian of transformation, defined as $G = \left[\det\left(G_{ij}\right)\right]^{\frac{1}{2}}$.
The covariant winds in this system are defined by

$$u = \mathbf{V} \cdot \mathbf{a}_1$$
$$v = \mathbf{V} \cdot \mathbf{a}_2.$$

(9.103)

The contravariant winds are related to the covariant winds via

$$\begin{pmatrix} \tilde{u} \\ \tilde{v} \end{pmatrix} = \mathbf{G}^{-1} \begin{pmatrix} u \\ v \end{pmatrix}$$

(9.104)

The relative vorticity is given by

$$\zeta = \frac{1}{G} \left(\frac{\partial v}{\partial x} - \frac{\partial u}{\partial y} \right)$$

(9.105)

and the kinetic energy by

$$K = \frac{1}{2} (\tilde{u} u + \tilde{v} v)$$

(9.106)

Further details on application of the general curvilinear coordinate system in the context of atmospheric primitive equations can be found in Sadourny (1972) and Rančić et al. (1996).

The momentum, the thermodynamic, and the continuity equations are respectively

$$\frac{\partial u}{\partial t} = (\zeta + f) G \tilde{v} - \frac{\partial K}{\partial x} - \dot{\eta} \frac{\partial u}{\partial \eta} - \frac{\partial \Phi}{\partial x} - \frac{RT_v}{p} \frac{\partial p}{\partial x} + F_u$$

(9.107)

$$\frac{\partial v}{\partial t} = -(\zeta + f) G \tilde{u} - \frac{\partial K}{\partial y} - \dot{\eta} \frac{\partial v}{\partial \eta} - \frac{\partial \Phi}{\partial y} - \frac{RT_v}{p} \frac{\partial p}{\partial y} + F_v$$

(9.108)

$$\frac{\partial T}{\partial t} = -\left(\tilde{u} \frac{\partial T}{\partial x} + \tilde{v} \frac{\partial T}{\partial y} \right) - \dot{\eta} \frac{\partial T}{\partial \eta} + \frac{1}{c_p} \frac{RT_v}{p} \omega + \frac{Q}{c_p}$$

(9.109)

$$\frac{\partial}{\partial \eta} \left(\frac{\partial p}{\partial t} \right) = -\frac{1}{G} \left[\frac{\partial}{\partial x} (\tilde{u} G m) + \frac{\partial}{\partial y} (\tilde{v} G m) \right] - \frac{\partial}{\partial \eta} (\dot{\eta} m)$$

(9.110)

The hydrostatic equation is written as

$$\frac{\partial \Phi}{\partial \eta} = -\frac{RT_v}{p} m$$

(9.111)

Here, the symbols have their usual meanings. The momentum equation is written in a vector invariant form, which conveniently encapsulates the curvilinear terms. The mass m is defined as

$$m = \frac{\partial p}{\partial \eta}$$

(9.112)

The vertical velocity $\dot{\eta}$ is derived from the continuity Eq. (9.110):

$$\dot{\eta} \frac{\partial p}{\partial \eta} = -\frac{\partial p}{\partial t} - \int_0^\eta \frac{1}{G} \left[\frac{\partial}{\partial x} (\tilde{u} G m) + \frac{\partial}{\partial y} (\tilde{v} G m) \right] d\eta$$

(9.113)

Similarly, the time tendency of surface pressure is computed by integrating the horizontal divergence over all layers:

$$\frac{\partial p_S}{\partial t} = - \int_0^{\eta_s} \frac{1}{G} \left[\frac{\partial}{\partial x} \left(\tilde{u} Gm \right) + \frac{\partial}{\partial y} \left(\bar{v} Gm \right) \right] d\eta \qquad (9.114)$$

In addition, the transport of a number of atmospheric scalar variables (specific humidity, cloud water, turbulent kinetic energy, etc.) is described by:

$$\frac{\partial q}{\partial t} = - \left(\bar{u} \frac{\partial q}{\partial x} + \bar{v} \frac{\partial q}{\partial y} \right) - \dot{\eta} \frac{\partial q}{\partial \eta} + Q_s \qquad (9.115)$$

where Q_s are various source and sink terms.

1.15 Microscale model (MIMO)

The numerical model MIMO is a three-dimensional model for simulating microscale window and dispersion of pollutants in built-up areas. It solves the Reynolds averaged conservation equations for mass, momentum and energy. Additional transport equations for humidity, liquid water content and passive pollutants can be solved. The Reynolds stresses and turbulent fluxes of scalar quantities can be calculated by several linear and nonlinear turbulence models.

The model MIMO solves the conservation equations for mass, momentum, energy and other scalar quantities such as the humidity or the concentration of passive pollutants. According to Reynolds (1895) the instantaneous value of a quantity $\tilde{\Phi}$ is split into a mean part Φ and a fluctuating part ϕ. In Table 1 the time averaged conservation equations for mass, momentum, energy, and other scalar quantities are given in a general form, which consists of terms describing the temporal derivative, the advection, the diffusion and source terms. U_i is the mean velocity and x_i are the Cartesian coordinates. The energy equation is formulated employing the potential temperature θ defined as

$$\theta = T \cdot \left(\frac{p_0}{p} \right)^{\frac{R_L}{c_p}} \qquad (9.116)$$

with the temperature T, the pressure p, and the reference pressure $p_0 (= 10^5 \text{ Pa})$ at ground level. The mean state is assumed to be in hydrostatic equilibrium, i.e., $\partial p / \partial z = -\rho g$. Usually the density fluctuations ρ' are small compared to the mean density ρ_0. Introducing the assumption $\rho' \ll \rho_0$ into the conservation equation for momentum yields the simplified equation

$$\frac{\partial (\rho_0 U_i)}{\partial t} + \frac{\partial (\rho_0 U_i U_j)}{\partial x_j} = -\rho' g_i - \frac{\partial p}{\partial x_i} - \frac{\partial (\rho_0 \overline{u_i u_j})}{\partial x_j} \qquad (9.117)$$

where density variations in the inertia term are neglected, but retained in the buoyancy term (Boussinesq approximation).

1.16 Regional atmospheric modeling system (RAMS)

The regional atmospheric modeling system is a comprehensive non-hydrostatic model developed at the Colorado State University (CSU), spearheaded by William R. Cotton and Roger A. Pielke, for mesoscale meteorological modeling. It is a set of computer programs that simulate the atmosphere for weather and climate research and for numerical weather prediction (NWP). The major components of RAMS include an atmospheric model which performs the actual simulations; a data analysis package which prepares initial data for the atmospheric model from observed meteorological data; and a post-processing model visualization and analysis package which interfaces atmospheric model output with a variety of visualization software utilities. In order to introduce a high degree of flexibility and versatility in RAMS, particularly regarding its new grid nesting capability, and to take advantage of the ever-increasing capabilities in computer hardware and software, RAMS was built on an entirely new framework, with the numerical schemes and parameterizations from the earlier models adapted to the new model structure.

The general equations for RAMS are described below. The equations are the standard hydrostatic or non-hydrostatic Reynolds-averaged primitive equations.

The non-hydrostatic equations are

i. Equations of motion,

$$\frac{\partial u}{\partial t} = -u\frac{\partial u}{\partial x} - v\frac{\partial u}{\partial y} - w\frac{\partial u}{\partial z} - \theta\frac{\partial \pi'}{\partial x} + fv + \frac{\partial}{\partial x}\left(K_m\frac{\partial u}{\partial x}\right) + \frac{\partial}{\partial y}\left(K_m\frac{\partial u}{\partial y}\right) + \frac{\partial}{\partial z}\left(K_m\frac{\partial u}{\partial z}\right)$$

$$\frac{\partial v}{\partial t} = -u\frac{\partial v}{\partial x} - v\frac{\partial v}{\partial y} - w\frac{\partial v}{\partial z} - \theta\frac{\partial \pi'}{\partial y} - fu + \frac{\partial}{\partial x}\left(K_m\frac{\partial v}{\partial x}\right) + \frac{\partial}{\partial y}\left(K_m\frac{\partial v}{\partial y}\right) + \frac{\partial}{\partial z}\left(K_m\frac{\partial v}{\partial z}\right)$$

$$\frac{\partial w}{\partial t} = -u\frac{\partial v}{\partial x} - v\frac{\partial w}{\partial y} - w\frac{\partial w}{\partial z} - \theta\frac{\partial \pi'}{\partial z} - \frac{g\theta'_v}{\theta_0} + \frac{\partial}{\partial x}\left(K_m\frac{\partial v}{\partial x}\right) + \frac{\partial}{\partial y}\left(K_m\frac{\partial v}{\partial y}\right) + \frac{\partial}{\partial z}\left(K_m\frac{\partial v}{\partial z}\right)$$

ii. Thermodynamic equation,

$$\frac{\partial \theta_{il}}{\partial t} = -u\frac{\partial \theta_{il}}{\partial x} - v\frac{\partial \theta_{il}}{\partial y} - w\frac{\partial \theta_{il}}{\partial z} + \frac{\partial}{\partial x}\left(K_h\frac{\partial \theta_{il}}{\partial x}\right) + \frac{\partial}{\partial y}\left(K_h\frac{\partial \theta_{il}}{\partial y}\right) + \frac{\partial}{\partial z}\left(K_h\frac{\partial \theta_{il}}{\partial z}\right) + \left(\frac{\partial \theta_{il}}{\partial t}\right)_{rad}$$

iii. Water species mixing ratio continuity equation, and

$$\frac{\partial r'_n}{\partial t} = -u\frac{\partial r_n}{\partial x} - v\frac{\partial r_n}{\partial y} - w\frac{\partial r_n}{\partial z} + \frac{\partial}{\partial x}\left(K_h\frac{\partial r_n}{\partial x}\right) + \frac{\partial}{\partial y}\left(K_h\frac{\partial r_n}{\partial y}\right) + \frac{\partial}{\partial z}\left(K_h\frac{\partial r_n}{\partial z}\right)$$

iv. Mass continuity equation.

$$\frac{\partial \pi'}{\partial t} = -\frac{R\pi_0}{c_v\rho_0\theta_0}\left(\frac{\partial \rho_0\theta_0 u}{\partial x} + \frac{\partial \rho_0\theta_0 v}{\partial y} + \frac{\partial \rho_0\theta_0 w}{\partial z}\right)$$

The hydrostatic equation in RAMS replaces the vertical equation of motion and the mass continuity equation with the hydrostatic equations.

RAMS is written primarily in Fortran with some C code and it runs best under the Unix operating system that has led to many improvements and new capabilities.

The capability of RAMS was recently augmented with the implementation of 2-way interactive grid nesting which are differential equations. In otherwords, with the understanding of numerical analysis of differential equation, it is possible for environmental researchers to combine models, formulate models, and expand models.

$$\frac{\partial \pi}{\partial z} = -\frac{g}{\theta_v} + g(r_T - r_v)$$

$$\frac{\partial \rho u}{\partial x} + \frac{\partial \rho v}{\partial y} + \frac{\partial \rho w}{\partial z} = 0$$

References

Agustí-Panareda, A., Beljaars, A., Cardinali, C., Genkova, I., Thorncroft, C., 2010. Impact of assimilating AMMA soundings on ECMWF analyses and forecasts. Weather Forecast. 25, 1142−1160.

Apicella, L., Puca, S., Lagasio, M., et al., 2021. The predictive capacity of the high resolution weather research and forecasting model: a year-long verification over Italy. Bull. of Atmos. Sci. Technol. 2, 3. https://doi.org/10.1007/s42865-021-00032-x.

Barthelet, P., Terray, L., Valcke, S., 1994. Transient CO_2 experiment using the ARPEGE/OPAICE non flux corrected coupled model. Geophys. Res. Lett. 25 (13), 2277−2280.

Bechtold, P., Köhler, M., Jung, T., Doblas-Reyes, F., Leutbecher, M., Rodwell, M.J., Vitart, F., Balsamo, G., 2008. Advances in simulating atmospheric variability with the ECMWF model: from synoptic to decadal time-scales. Q. J. R. Meteorol. Soc. 134, 1337−1351.

Bibiana, S.C.da C., da Rocha, N.S., da Silva, S.C.S., et al., 2018. The use of Weather Research and Forecasting Model to estimate the vertical profile of meteorological data. IEEE Int. Symp. Geosci. Remote Sens. IGARSS 5560−5563.

Bush, M., Allen, T., Bain, C., Boutle, I., Edwards, J., Finnenkoetter, A., Franklin, C., Hanley, K., Lean, H., Lock, A., Manners, J., Mittermaier, M., Morcrette, C., North, R., Petch, J., Short, C., Vosper, S., Walters, D., Webster, S., Weeks, M., Wilkinson, J., Wood, N., Zerroukat, M., 2019. The first met office unified model/JULES regional atmosphere and land configuration, RAL1. Geosci. Model Dev. (GMD) 13, 1999−2029. https://doi.org/10.5194/gmd-2019-130.

Brown, A., Milton, S., Cullen, M., Golding, B., Mitchell, J., Shelly, A., 2012. Unified modeling and prediction of weather and climate: a 25 year journey. Bull. Am. Meteorol. Soc. 93, 1865−1877. https://doi.org/10.1175/BAMS-D-12-00018.1.

Chen, F., Janjic, Z., Mitchell, K., 1997. Impact of atmospheric surface-layer parameterizations in the new land-surface scheme of the NCEP mesoscale Eta Model. Bound Layer Meteorol. 85, 391−421.

Chuang, H.-Y., Manikin, G., 2001. The NCEP Eta Model Post Processor: A Documentation. NCEP Office Note 438, 52 pp. http://www.emc.ncep.noaa.gov/officenotes/FullTOC.html#2000.

Colle, B.A., Olson, J.B., Tongue, J.S., 2003. Multiseason verification of the MM5. Part I: comparison with the Eta Model over the central and eastern United States and impact of MM5 resolution. Weather Forecast. 18, 431−457.

Côté, J., Desmarais, J.-G., Gravel, S., Méthot, A., Patoine, A., Roch, M., Staniforth, A., 1998. The operational CMC-MRB global environmental Multiscale (GEM) model: Part II - results. Mon. Weather Rev. 126, 1397–1418.

Coy, L., Allen, D.R., Eckermann, S.D., McCormack, J.P., Stajner, I., Hogan, T.F., 2007. Effects of model chemistry and data biases on stratospheric ozone assimilation. Atmos. Chem. Phys. 7, 2917.

de Rooy, W.C., 2014. The Fog above Sea Problem: Part 1 Analysis. ALADIN-HIRLAM Newsletter, No. 2. Météo-France, Centre National de Recerches Meteorologiques, Toulouse, France, pp. 9–15. http://hirlam.org/index.php/component/docman/doc_download/1490-aladin-hirlam-newsletter-no-2-april-2014?Itemid=218.

Déqué, M., Dreveton, C., Braun, A., et al., 1994. The ARPEGE/IFS atmosphere model: a contribution to the French community climate modelling. Clim. Dynam. 10, 249–266. https://doi.org/10.1007/BF00208992.

Dutton, J.A., 1986. The Ceaseless Wind, An Introduction to the Theory of Atmospheric Motion. McGraw-Hill, New York, p. 617.

Eckermann, S.,D., McCormack, J.P., Coy, L., Allen, D., Hogan, T., Kim, Y.-J., 2004. NOGAPS-ALPHA: A Prototype High-Altitude Global NWP Model, Anniversary of Operational Numerical Weather Prediction. American Meteorological Society, pp. 14–17.

Eckermann, S.D., Wu, D.L., Doyle, J.D., Burris, J.F., McGee, T.J., Hostetler, C.A., Coy, L., Lawrence, B.N., Stephens, A., McCormack, J.P., Hogan, T.F., 2006. Imaging gravity waves in lower stratospheric AMSU-A radiances, Part 2: validation case study. Atmos. Chem. Phys. 6, 3343.

Forbes, R.M., Ahlgrimm, M., 2014. On the representation of high-latitude boundary-layer mixed-phase cloud in the ECMWF global model. Mon. Weather Rev. 142, 3425–3445.

Fujita, Ts., 2003. Higher order finite difference schemes for advection of NHM. Proceedings, CAS/JSC WGNE Res. Act. atmosphere ocean Model. 33, 3.09–3.10.

Gao, J., Stensrud, D.J., 2012. Assimilation of reflectivity data in a convective-scale, cycled 3DVAR framework with hydrometeor classification. J. Atmos. Sci. 69, 1054–1065.

Gill, A.E., 1982. Atmosphere-Ocean Dynamics. Academic Press, New York, p. 662.

Gleeson, E., Nielsen, K., Toll, V., Rontu, L., 2015. Shortwave Radiation Experiments in HARMONIE. Tests of the Cloud Inhomogeneity Factor and a New Cloud Liquid Optical Property Scheme Compared to Observations. ALADIN-HIRLAM Newsletter, No. 5. Météo-France, Centre National de Recerches Meteorologiques, Toulouse, France, pp. 92–106. http://www.umr-cnrm.fr/aladin/IMG/pdf/nl5.pdf.

Haltiner, G.J., Williams, R.T., 1980. Numerical Prediction and Dynamic Meteorology. John-Wiley and Sons, p. 477.

Ikawa, M., Saito, K., 1991. Description of a non-hydrostatic model developed at the forecast research department of the MRI. MRI Tech. Rep. 28, 238.

Kumagai, Y., 2004. Implementation of a non-local like PBL scheme in JMANHM. CAS/JSC WGNE Res. Activ. Atmos. Oceanic Modell. 34, 0417–0418.

Leonard, S., Turner, J., Milton, S., 1997. An assessment of UK Meteorological Office numerical weather prediction analyses and forecasts for the Antarctic. Antarct. Sci. 9 (1), 100–109. https://doi.org/10.1017/S0954102097000126.

Malcolm, A., Glover, M., Selwood, P., 2010. Scalability of the Met Office Unified Model, 14th ECMWF workshop. https://www.ecmwf.int/node/15065.

McCormack, J.P., Eckermann, S.D., Coy, L., Allen, D.R., Kim, Y.-J., et al., 2004. NOGAPS-ALPHA model simulations of stratospheric ozone during the SOLVE2 campaign. Atmos. Chem. Phys. 4, 2401–2423. https://doi.org/10.5194/acp-4-2401-2004.

Mesinger, F., 1996. Improvements in quantitative precipitation forecasts with the Eta regional model at the National Centers for Environmental Prediction. The 48-km upgrade. Bull. Am. Meteorol. Soc. 77, 2637–2649.

Mesinger, F., Chou, S.C., Gomes, J.L., et al., 2012. An upgraded version of the Eta model. Meteorol. Atmos. Phys. 116, 63–79. https://doi.org/10.1007/s00703-012-0182-z.

Mesinger, F., 1984. A blocking technique for representation of mountains in atmospheric models. Riv. Meteorol. Aeronaut. 44, 195–202.

Mesinger, F., Janjić, Z.I., 1985. Problems and numerical methods of the incorporation of mountains in atmospheric models. In Large-scale Com- putations in Fluid Mechnaics, Part 2. Lect. Appl. Math. 22, 81–120.

Michalakes, J.G., Dudhia, J., Gill, D.O., Klemp, J.B., Skamarock, W.C., 1999. Design of a next-generation regional weather research and forecast model. In: Zweiflhofer, W., Kreitz, N. (Eds.), Towards Teracomputing: Proceedings of the Eighth ECMWF Workshop on the Use of Parallel Processors in Meteorology. World Scientific, pp. 117–124.

Mukhopadhyay, P., Prasad, V.S., Phani Murali Krishna, R., Deshpande, M., Ganai, M., Tirkey, S., Sarkar, S., Goswami, T., Johny, C.J., Roy, K., Mahakur, M., Durai, V.R., Rajeevan, M., 2019. Performance of a very high-resolution global forecast system model (GFS T1534) at 12.5 km over the Indian region during the 2016–2017 monsoon seasons. J. Earth Syst. Sci. 128, 155. https://doi.org/10.1007/s12040-019-1186-6.

NOAA, 2022. Global Forecast System. https://www.ncei.noaa.gov/products/weather-climate-models/global-forecast.

Ooyama, K.V., 1990. A thermodynamic foundation for modeling the moist atmosphere. J. Atmos. Sci. 47, 2580–2593.

Patel, R.N., Yuter, S.E., Miller, M.A., Rhodes, S.R., Bain, L., Peele, T.W., 2021. The diurnal cycle of winter season temperature errors in the operational global forecast system (GFS). Geophys. Res. Lett. 48 (20), 1–9. https://doi.org/10.1029/2021GL095101.

Peckham, S.E., Grell, G.A., McKeen, S.A., Schmitz, R., 2004. Comparisons between observations made during NEAQS and air quality forecasts from MM5 and WRF chemistry models. In: Preprints, Sixth Conf. on Atmospheric Chemistry: Air Quality in Megacities. Amer. Meteor. Soc., CD-ROM, J2.15, Seattle, WA.

Phillips, N.A., 1966. The equations of motion for a shallow rotating atmosphere and the "traditional approximation". J. Atmos. Sci. 23, 626–629.

PSU/NCAR, 2005. Introduction to MM5 Modeling System. https://www2.mmm.ucar.edu/mm5/documents/MM5_tut_Web_notes/INTRO/intro.htm.

Purser, R.J., Rančić, M., 1997. Conformal octagon: an attractive framework for global models offering quasiuniform regional enhancement of resolution. Meteorol. Atmos. Phys. 62, 33–48.

Rančić, M., Purser, R.J., Mesinger, F., 1996. A global shallow- water model using an expanded spherical cube: Gnomonicversus conformal coordinates. Quart. J. Roy. Meteor. Soc. 122, 959–982.

Reynolds, O., 1895. On the Dynamical theory of Incompressible Viscous Fluids and the Determination of the Criterion. Phil. Trans. Roy. Soc. (London) 186A, 123–164.

Ritchie, H., 1997. Application of the semi-Lagrangian method to global spectral forecast models. Atmos.-Ocean 35, 445–467. https://doi.org/10.1080/07055900.1997.9687360.

Rivest, C., Staniforth, A., Robert, A., 1994. Spurious resonant response of semi-Lagrangian discretizations to orographic forcing: diagnosis and solution. Mon. Weather Rev. 122, 366–376.

Sadourny, R., 1972. Conservative finite-difference approximations of the primitive equations on quasi-uniform spherical grids. Mon. Wea. Rev. 100, 136–144.

Saito, K., Yamada, Y., Fujita, T., Ishida, J., Tanaka, S., 2004. The JMANHM started its pre-operational daily run (in Japanese). NWP News 23 (1), 1–11. Japan Meteorological Agency.

Senkova, A., Rontu, L., Savijärvi, H., 2007. Parametrization of orographic effects on surface radiation in HIRLAM. Tellus 59A, 279–291. https://doi.org/10.1111/j.1600-0870.2007.00235.x.

Strazdins, P.E., Kahn, M., Henrichs, J., Pugh, T., Rezny, M., 2011. Profiling methodology and performance tuning of the met office unified model for weather and climate simulations. In: Parallel and Distributed Processing Workshops and Phd Forum (IPDPSW), 2011 IEEE International Symposium on Parallel and Distributed Processing Workshops and Phd Forum, pp. 1322–1331. https://doi.org/10.1109/IPDPS.2011.283.

Simmons, A.J., Burridge, D.M., 1981. An energy and angular-momentum conserving vertical finite difference scheme and hybrid vertical coordinates. Mon. Wea. Rev. 109, 758–766.

Simmons, A.J., Strüfing, R., 1981. An energy and angular-momentum conserving finite difference scheme, hybrid coordinates and medium-range weather prediction. In: ECMWF Tech. Report No., 28 European Centre for Medium-Range Weather Forecasts, Reading, England, p. 68.

Smyth, S., Yin, D., Roth, H., Jiang, W., 2005. A Study of the Impact of GEM and MM5 Meteorology on CMAQ Modelling Results in Eastern Canada and the Northeastern United States. Institute for Chemical Process and Environmental Technology Tech. Rep. PET-1561-04S, p. 77.

United Nations, 2022. European Centre for Medium-Range Weather Forecasts (ECMWF). https://un-spider.org/european-centre-medium-range-weather-forecasts-ecmwf.

von Bezold, W., 1888. On the thermo-dynamics of the atmosphere (first communication). In: The Mechanics of the Earth Atmosphere, a collection of translations by Cleveland Abbe, 1891. Smithsonian Miscellaneous Collections. [Translated from the Proceedings of the Royal Prus- sian Academy of Science at Berlin (Sitzungsberichte der König. Preuss. Akademie der Wissenschaften zu Berlin), pp. 485–522. Art. XV, 212–242.

Wacker, U., Herbert, F., 2003. Continuity equations as expressions for local balances of masses in cloudy air. Tellus 55A, 247–254.

Xue, M., Droegemeier, K.K., Wong, V., Shapiro, A., Brewster, K., 1995. ARPS Version 4.0 User's Guide. Center for Analysis and Prediction of Storms [Available from: CAPS, Univ. of Oklahoma, 100 E. Boyd St., Norman OK 73019], 380 pp.

Xue, M., Wang, D., Gao, J., Brewster, K.A., Droegemeier, K., 2003. The Advanced Regional Prediction System (ARPS), storm-scale numerical weather prediction and data assimilation. Meteorol. Atmos. Phys. 82, 139–170.

Yamada, Y., 2003. Cloud microphysics. The JMA nonhydrostatic model. Japan Meteorol. Agency Annual Rep 49, 52–76.

Yamazaki, Y., Saito, K., 2004. Implementation of the cylindrical equidistant projection for the non-hydrostatic model of the Japan Meteorological Agency. CAS/JSC WGNE Res. Activ. Atmos. Oceanic Modell. 34, 0327–0328.

Yáñez-Morroni, G., Gironás, J., Caneo, M., Delgado, R., Garreaud, R., 2018. Using the weather research and forecasting (WRF) model for precipitation forecasting in an andean region with complex topography. Atmosphere 9 (8), 304. https://doi.org/10.3390/atmos9080304.

Yeh, K.-S., Côté, J., Gravel, S., Méthot, A., Patoine, A., Roch, M., Staniforth, A., 2002. The CMC—MRB Global Environmental Multiscale (GEM) model. Part III: nonhydrostatic formulation. Mon. Weather Rev. 130, 356.

Yue, H., Gebremichael, M., Nourani, V., 2022. Evaluation of Global Forecast System (GFS) medium-range precipitation forecasts in the Nile River Basin. J. Hydrometeorol. 23 (1), 101—116. https://doi.org/10.1175/JHM-D-21-0110.1.

Saito, K., 1997. Semi-implicit fully compressible version of the MRI mesoscale nonhydrostatic model —Forecast experiment of the 6 August 1993 Kagoshima torrential rain—. Geophys. Mag. Ser. 2, 109—137.

Saito, K., Kato, T., Eito, H., Muroi, C., 2001. Documentation of the Meteorological Research Institute/Numerical Prediction Division Unified Nonhydrostatic Model. Tec. Rep. MRI 42, 133.

Further reading

Fujita, T., 2003. Higher order finite difference schemes for advection of NHM. In: Proc. Int. Workshop on NWP Models for Heavy Precipitation in Asia and Pacific Areas. Japan Meteorological Agency, Tokyo, Japan, pp. 78—81.

Furevik, T., Bentsen, M., Drange, H., Kindem, I., Kvamstø, N.G., Sorteberg, A., 2003. Description and evaluation of the bergen climate model: ARPEGE coupled with MICOM. Clim. Dynam. 21, 27—51.

Raghavan, S.V., Vu, M.T., Liong, S.Y., 2015. Regional climate simulations over Vietnam using the WRF model. Theor. Appl. Climatol. 126 (1—2), 1—10. https://doi.org/10.1007/s00704-015-1557-0.

Index

'*Note:* Page numbers followed by "f" indicate figures and "t" indicate tables.'

A

Additivity, 82
Adomian decomposition method (ADM), 129–130
Advanced Region Prediction System (ARPS)
  Cartesian coordinate, 206
  equations of state, 206
Advanced Very High Resolution Radiometer (AVHRR), 60–61
Aerosol optical depth (AOD), 131–132
  Basse-Gambia, 134f
  dataset, 132
  Kanpur-India, 134f
Aire Limitée Adaptation dynamique Développement InterNational (ALADINs)
  continuity equation, 211–212
  gas constants and specific heats, 211
  hydrostatic equation, 210
  latent heats, 211
  moisture equation, 211
  momentum equation, 210
  pressure-based hybrid coordinate, 210
  surface pressure equation, 212
  thermodynamic equation, 210
Air pollution
  bioaerosol production, 20–21
  causes, 17
  control, 18
  definition, 14
  effects, 17–18
  fossil fuel burning, 14–15, 15f
  garbage burning, 17
  gas flaring, 19–20
  model formulation diagram, 17
  transportation, 16–17
  vehicular pollution, 18–19
  wildfire, 15
Analytica, 7
Angstrom exponent, 134–135
Approximation error, 151
Artificial Neural Network (ANN) approach, 171

B

Batteries, electronic waste pollution, 35
Bicubic interpolation, 169–170
Bioaerosol production, 20–21
  bioaerosols, 21

  cancer, 21
  harmful effects, 21
  living and nonliving things, 20–21
  transmittable/transferrable diseases, 21
Bioremediation, soil contamination, 24
Bisection method, 81–82, 92–103
  computational application, 112–117
  homogeneity, 92–103
Bore-hole pollution
  contamination, 30f
  landfills, 30
  man-made products, 29
  prevention, 31
  septic system, 29
  storage tanks, 30

C

Cadmium, electronic waste pollution, 37–38
Cancer, bioaerosol production, 21
Carbon dioxide emission, 15f
Chemical pollution, 24
Chromium, electronic waste pollution, 37–38
Climate change
  drought, 50
  emissions and overexploitation, 47–49
  greenhouse gases, 47–49, 49f
  Greenland and Antarctic ice sheets, 50
  temperature, 47–49
  weather patterns, 50
  wildfire risk and severity, 49
Climate data record (CDR), 60, 61f
Communication errors, 54–55
Computational technique, 7
Conditional probability, 10
Contingent databases, 2–3
Continuous data, 1–2
Continuous simulation (CS) models, 47
Conversion error, 56
Coverage error, 56

D

Data
  definition, 1
  measurement technique, 4
  multiple trials, 4
  primary, 1, 3–4
  qualitative, 1–2

Data (*Continued*)
  quantitative, 1—2
  secondary, 1, 4
  source, 1—2
Databases, 2—3
Data Collection Platforms (DCP), 60
Data noise, 4
Data quality and errors, 55—60
  background survey information, 59t
  comparability of statistics, 55
  conversion error, 56
  coverage error, 56
  measurement error, 56—57
  nonresponse error, 56
  numerical errors, 57
  processing error, 57
  sampling error, 55—56
  Taylor's theorem, 57
Data sampling, 9
Datasets, 53
Data treatment
  computational technique, 7
  data collection, 5—6
  data noise, 4
  data preparation, 5—6
  data processing, 5—6
  mathematical technique, 6—7
  sample errors, 5
  statistical data treatment, 7—11
Delaware Estuary Comprehensive Study (DECS)
    model, 25
Discrete data, 1—2
Dispersion models, 68—69
Distribution range, 10

E

Electronic waste pollution, 35—38
Environmental/atmospheric numerical models
  Advanced Region Prediction System (ARPS),
      206—207
  Aire Limitée Adaptation dynamique Développe-
      ment InterNational (ALADINs), 210—212
  Eta model, 213—215
  European Center for Medium Range Weather
      Forecasts (ECMWF), 196—197
  fifth generation mesoscale model, 205—206
  French global atmospheric forecast model
      (ARPEGE), 199—200
  Global Environmental Multiscale limited area
      model, 208—210
  Global Environmental Multiscale Model (GEM),
      195

global forecast system (GFS), 191—192
  High Resolution Limited Area Model (HIR-
      LAM), 207—208
  Japan Meteorological Agency Nonhydrostatic
      Model (JMA-NHM), 203—204
  microscale model (MIMO), 215
  NOGAPS-ALPHA model, 192—195
  regional atmospheric modeling system (RAMS),
      216—217
  Unified Model (UKMO), 197—199
  Weather Research and Forecasting (WRF) model,
      200—202
Environmental concerns
  climate change, 47—50
  rainfall, 44—47
  thermal comfort, 41—43
Environmental Data Record (EDR), 60
Environmental law enforcement data, 55
Environment databases, 2—3
Error detection, 54—55
Eta model, 213—215
  atmospheric scalar variables, 215
  blocking effect, 213
  continuity equation, 214
  contravariant winds, 214
  covariant winds, 213—214
  hydrostatic equation, 214
  hydrostatic vertical eta (eta) coordinate, 213
  kinetic energy, 214
  mass, 214
  metric tensor of transformation, 213
  momentum equation, 214
  relative vorticity, 214
  thermodynamic equation, 214
  time tendency, surface pressure, 215
  vertical velocity, 214
Eulerian modeling, 68
Euler method
  application of, 121
  computation processing, 136—142
  error, 121
  function, 121
  improved, 122—123
  initial condition, 122
  Taylor's series, 121
  theory, 121
European Center for Medium Range Weather
    Forecasts (ECMWF)
  continuity equation, 196
  finite-element based discretization, 196
  humidity equation, 196
  hybrid vertical coordinate, 196—197

hydrostatic equation, 196
momentum equation, 196
semi-Lagrangian advection scheme, 197
thermodynamic equation, 196
Event-based (EB) model, 47
e-waste, 36, 36t
Excavation, soil contamination, 24
Experimental procedure
experimental safety rules, 73
general safety rules, 74
post-experimental safety, 73–74
pre-experimental safety, 70–73
safety rules, 69–70

F
Factor analysis, 6
Fifth generation mesoscale model
advection terms, 205
divergence term, 205
momentum equation, 205
pressure equation, 205
thermodynamics equation, 205
Flat files, 2
FlexPro, 7
Fossil fuel consumption, 15f
Fractional derivative approach, 129–130
Framework for the Development of Environment
Statistics (FDES), 53–54
FreeMat, 7
French global atmospheric forecast model
(ARPEGE)
continuity equation, 199
molecular momentum transport, 199
momentum equations, 199
self-diffusive transport of mass, 200
Frequency analysis, rainfall data, 44
Fundamental climate data record (FCDR), 60

G
Garbage burning, 17
Gas flaring, 19–20, 20f
Gaussian plume model, 68–69
Geographic context, environmental data, 53
Global Environmental Multiscale limited area
model
hybrid coordinate, 209
vertical coordinate, 209–210
Global Environmental Multiscale Model (GEM)
shallow atmosphere approximation, 195
spherical coordinate, 195
Global forecast system (GFS)
density, 191–192

Euler equation, 191–192
high-resolution global forecast system model, 192
Nile river basin, medium-range precipitation
forecasts, 192
Global Observing System (GOS), 60
Global Precipitation Measurement (GPM)
constellations, 3f
Greenhouse effect, 17
Groundwater pollution, 24

H
Heun's method, 126–128
High-resolution global forecast system model, 192
High Resolution Limited Area Model (HIRLAM)
continuity equation, 208
divergence operator, 208
forecast model equations, 207
hybrid system, 207
hydrostatic equation, 208
momentum equations, 207–208
pressure vertical velocity equation, 208
spherical rotated coordinates, 207
surface pressure tendency, 208
thermodynamic and moisture equations,
207–208
Homogeneity, 83–103
Hyetograph method, 44

I
Improved Euler method
average values, 122–123
calculation, 123
initial condition, 123
Interpolation
computational application, 181–188
environmental data
air quality data, 172
environmental air pollution, 171
Lagrange interpolation, 172–175
Newton interpolation, 176–178
optimal sampling strategy, 172
satellite observations and emission inventories,
170
spatial interpolation, 171
Spline interpolation, 179–181
water quality assessment, 171
formulation, 169
importance, 169
polynomial route, 169
types, 169–170
Inverse Distance Weighted (IDW) interpolation,
171

J

Japanese space program, 63, 64t
Japan Meteorological Agency Nonhydrostatic
   Model (JMA-NHM)
   continuity equation, 204
   density, 204
   Lambert conformal projection, 203
   mass-virtual potential temperature, 204
   spherical curvilinear orthogonal coordinate, 203
   state equation, 204
   thermodynamic equation, 204
jLab, 7
Joint Center for Satellite Data Assimilation
   (JCSDA), 63

L

Laboratory apparatus maintenance, 75–76
Laboratory practice errors, 74–75
Lagrange interpolation
   advection-diffusion-reaction, 173
   data point, 174
   first order interpolation, 173–174
   mobile sprinkler machine, water distribution
      calculation, 173
   Newton–Cotes method, 172–173
   Phenol Groundwater Transport (PGWT) equa-
      tion, 173
   Vandermonds determinants, 175
Lagrangian modeling, 68
Landfills, bore-hole pollution, 30
Land pollution
   causes, 22
   control measures, 22
   effects, 22
   soil contamination, 23–24
Lead, electronic waste pollution, 37–38
Linear interpolation, 169–170
Linear regression, 11

M

Mass curve, rainfall method, 44
Mathematical technique, 6–7
Mathematics modeling, water environment, 68f
MATLAB, 7
Mean, 8
Measurement error, 56–57, 75
Median, 9
Mercury, electronic waste pollution,
   37–38
Meteorological models, 68–69
Microbial pollution, 24
Microscale model (MIMO), 215

Midpoint method
   corrector step, 128
   predictor step, 128
Midpoint rule
   approximation error, 151
   error bound, 151
   relative error, 151
   Riemann sum, 149
   subintervals, 149–150
Mode, 8–9
Modeling procedure
   calibration and verification phase, 66f
   conceptualization stage, 63–66
   diagnostic models, 68–69
   Eulerian modeling, 68
   Lagrangian modeling, 68
   mass conservation principle, 68
   meteorological models, 68–69
   one-dimensional push-flow model, 68
   priori and posteriori knowledge, 65f
   research/management models, 66–67
Multiple linear regression, 11
Multispectral satellite missions, 61, 62t

N

National Oceanic and Atmospheric Administration (NOAA) missions, 62
Naturally occurring radioactive material (NORM), 34t
Newton-Cotes method, 172–173
Newton interpolation
   backward divided difference, 178
   divided difference interpolation polynomial, 177
   divided differences, 176
   forward difference, 177
   Newton's polynomial, 176
Newton's method, 80
   computational application, 104–109
   homogeneity, 86–88
NOGAPS-ALPHA model
   continuity equation, 193
   moisture conservation equation, 193–194
   photochemistry parameter forecasting, 192
   pressure, 192–193
   tendency equation, 193
   thermodynamic energy equation, 193
   vertical coordinate, 192–193
   vorticity and divergence, 194
Noise pollution
   effects, 32
   sources, 32
Nonresponse error, 56
Numerical data, 1–2

Numerical differential analysis
  computational application, 142
  computation processing, 136—142
  environmental research, lake pollution model
    Adomian decomposition method (ADM),
      129—130
    aerosol optical depth (AOD) dataset, 132—133
    aerosol size distribution, 133
    air pollution, 130—132
    air quality, 130
    Angstrom exponent, 134—135
    Basse location, 132, 132f
    formulation, 129
    fractional derivative approach, 129—130
    fractional differential equations, 129
    pollutant, 129, 131
    pollutant dynamics, 129—130
    satellite measurement, 133
    semianalytic technique, 130
    three-dimensional system, 130
  first-order homogeneous linear differential equa-
    tion, 119
  ordinary differential equation
    Euler method, 121—122
    improved Euler method, 122—123
    midpoint method, 128
    Predictor Corrector Method, 126—128
    Runge—Kutta method, 123—126
Numerical errors, 57
Numerical integration application
  computational application, 158—168
  midpoint rule
    approximation error, 151
    error bound, 151
    relative error, 151
    Riemann sum, 149
    subintervals, 149—150
  Simpson's rule
    ab-initio quadratic function, 156
    absolute error, 158
    error bound, 157
    quadratic function, 155
    relative error, 158
    Simpson's 1/3 rule, 154
    subintervals, 157
  trapezoidal rule, 151—154
    absolute error, 153
    area under the curve, 151—152
    error bound, 154
    numerical integration, 152
    relative error, 153
    subinterval, 152—153

Numerical meteorological models, 68—69
Numerical weather prediction (NWP), 63
Nutrient pollution, 24

O

Observational error, 75
Observation metrological model,
  68—69
Oil spillage, 32
One-dimensional push-flow model, 68
One-dimensional tropical hydrodynamic model,
  63—66
Online media information, 3
Operational Land Imager (OLI), 61—62
Oxygen depletion pollution, 24

P

Parallax error, 75
Pathfinder Atmospheres-Extended (PATMOS-x),
  60—61
Percentage of people dissatisfied (PPD) method,
  41—42
Phenol Groundwater Transport (PGWT) equation,
  173
Piecewise constant interpolation, 169—170
Point rainfall method, 45
Political pressing factor, 54
Pollutants, 13
Pollution
  air, 14—21
  definition, 13
  electronic waste, 35—38
  land, 21—24
  noise, 32
  radioactive, 33—34
  types, 13
  water pollution, 24—32
Polynomial interpolation, 170
Predictor Corrector Method
  advantages, 126—127
  algorithms, 126
  calculation, 127
  corrector step, 127
  initial condition, 127
  predictor step, 126
  second predictor step, 127
  second time step, 127
Pre-experimental safety, 70—73
Primary data, 1
Processing error, 57

Q

Qualitative data research, 1–2
Qualitative environmental data, 53
Quantitative data research, 1–2

R

Radioactive pollution
  causes, 34
  definition, 33
  effects, 34
  naturally occurring radioactive material
      (NORM), 33, 34t
  radionuclides, rock, 33t
  radon emissions, 34, 35t
  uranium and radium, 34
Rainfall–runoff (RR) models, 47
Rainfall system
  analyzing methods, 44–45
  challenges, 44–45
  changing atmosphere, 45–46
  continuous simulation (CS) models, 47
  event-based (EB) model, 47
  goodness-of-fit measures, 48t
  lagos, 46–47
  rainfall–runoff (RR) models, 47
  rain measurements, 44
  rain patterns, 46f
  spatial distribution, 47
  weather systems, 45
Random errors, 5, 74–75
Range, 9
Reforestation, soil contamination, 24
Regional atmospheric modeling system (RAMS)
  components, 216
  hydrostatic equation, 216–217
  mass continuity equation, 216
  motion equation, 216
  thermodynamic equation, 216
  water species mixing ratio continuity equation, 216
Regression, 6, 10
River contamination, 31–32, 31f
Root finding technique, 79
  additivity, 82
  Bisection method, 81–82
  computational application
    bisection method, 112–117
    Code5.8, 116–117
    Newton's method, 104–109
    secant method, 109–112
  homogeneity, 83–103
    bisection method, 92–103
    Microsoft Excel Package, 83–103

  Newton's method, 86–88
    radioactive dose measurement, 83t
    secant method, 89–92
    thorium measurement, 86
  Newton's method, 80
  nonlinear systems, 82
  roundoff errors, 82
  Secant method, 80–81
  truncation errors, 82
Roundoff errors, 82
Runge-Kutta method
  calculation, 124
  explicit methods, 123–124
  implicit methods, 123–124
  initial condition, 124–125
  initial value problem, 123–124
  iterative process, 125
  second-order, 125
  third-order, 125
Runge phenomenon, 179

S

Safety rules
  environmental field researcher, 69–70
  environmental laboratory researcher, 70
  field research, 70, 71t–72t
Sampling error, 55–56
Satellite measurement
  climate data record (CDR), 60, 61f
  cross- and inter-satellite calibrations, 60–61
  experimental missions, 63
  Landsat 8, 61–62
  multispectral satellite missions, 61, 62t
  National Oceanic and Atmospheric Administration (NOAA) missions, 62
  numerical weather prediction (NWP), 63
Scaled Lagged Averaged Forecasting (SLAF)
    method, 206
Scale error, 75
Secant method, 80–81, 89–92
  computational application, 109–112
  homogeneity, 89–92
Secondary data, 1, 4
Septic system, bore-hole pollution, 29
Simpson's rule
  ab-initio quadratic function, 156
  absolute error, 158
  error bound, 157
  quadratic function, 155
  relative error, 158
  Simpson's 1/3 rule, 154
  subintervals, 157

Smog, 18
Soil contamination, 23—24, 23f
  agricultural practices, 24
  chemical pollutants, 23
  hazardous substance deposition, 23
  human activities, 23
  nonagricultural sources, 24
  soil pollution, 23
Spatial interpolation, 171
Spline interpolation
  and Cressman interpolations, 181
  graphic user interface software, 179
  linear splines, 179—180
  piecewise continuous function, 180
  quadratic splines, 179
  Runge phenomenon, 179
  spline, 180
  spline formular, 180
  spline functions, 179
  spline smooth 2n parameter, 180
Spontaneous errors, 5
Standard deviation, 9
Statistical data treatment, 7—11
Statistics, 53—54
Stochastic models, 68—69
Storage tanks, bore-hole pollution, 30
Surface water pollution, 24
Suspended matter pollution, 24
Systematic errors, 5, 75

T
Technically enhanced naturally occurring radio-
    active material (TE-NORM), 34t
Thermal comfort
  across continents, 43f
  in buildings, 41
  control analysis method, 43
  definition, 41
  humidity, 41
  illustration, 42f
  parameters, 42—43
  percentage of people dissatisfied (PPD) method,
    41—42
  predicted mean vote (PVM) method, 41—42
Thermal remediation, soil contamination, 24
Time series analysis, 6
Trapezoidal rule, 151—154
  absolute error, 153
  area under the curve, 151—152
  error bound, 154

  numerical integration, 152
  relative error, 153
  subinterval, 152—153
Truncation errors, 82

U
Unified Model (UKMO)
  compressible Euler equations, 198
  fully compressible equations, 198
  limited accessibility, 197—198
  merits and demerits, 198—199
  thermodynamic variables, 198

V
Variance, 11
Vehicular pollution, 18—19, 18f

W
Water pollution
  bore-hole pollution, 29—31
  causes, 25
  control measures, 26
  effects, 25
  river contamination, 31—32
  types, 24
  water research data models, 25
  well pollution, 26—29
Weather Research and Forecasting (WRF) model
  actual atmospheric conditions, 200
  dry-air mass, vertical coordinate, 201—202
  general prognostic equations, 200
  inverse density, 200—201
  NavierStokes equations, 200—201
  vertical coordinate, 200
  vertical variable, 200—201
Web Services, 2
Well pollution, 26—29
  causes, 28
  contaminated well, 26f
  dangers, 28—29
  eutrophication, 27—28
  groundwater wells, 26—27
  mathematical model, 27
  policy, 29
  water quality models, 27
  water regulations, 29
Whittaker—Shannon interpolation, 170
Wildfire, 15, 16f

Printed in the United States
by Baker & Taylor Publisher Services